An Editor's Guide to Writing and Publishing Science

ENDORSEMENTS "An Editor's Guide to Writing and Publishing Science"

"Like the mythological Labyrinth, the contemporary world of scientific publishing can appear tortuous and even terrifying, particularly to the uninitiated. Fortunately, Michael Hochberg—respected population biologist, longtime editor of *Ecology Letters*, and latter-day Daedalus—knows this labyrinth inside and out. In his lively new book, Hochberg provides a unique and eminently readable guide to navigating every imaginable step of the process. This book will be useful both in easing the path of early career professionals through the publishing experience and in acquainting even experienced editors with the latest twists and turns in the business."

May Berenbaum, Professor, University of Illinois at Urbana-Champaign, and Editor in Chief of *PNAS*

"I've never seen a book on scientific writing like this. Of course, Hochberg covers how to write a title, an abstract, the methods, the results, a discussion section. But he also makes a deep dive into the workings of the journal publishing process. He teaches the reader how to navigate the rules and conventions—formal and informal, written and unwritten—that govern these institutions of scholarly communication."

Carl Bergstrom, Professor, University of Washington, and co-author of forthcoming book *Calling Bullshit*

"A scientific publication, when it appears in a journal, is a product that tells very little about the process that led to its existence. Young scientists are lucky if they have someone to tell them what happens behind the scenes. The author of this book has extensive experience of what really happens in scientific publishing, and luckily for all of us, he is willing to lift the curtain for everyone to have a look."

Hanna Kokko, Professor, University of Zurich, and author of *Modelling for Field Biologists and Other Interesting People*

"Michael Hochberg's book is a perfect guide for authors, because he is acutely aware of what it means to be on either side of the current journal-based system by which research and researchers are evaluated. His book, however, is much more than that. It is also testament to an integrity in science that is no longer prioritized, precisely because of the perverse incentives created around where and not what you publish. Publishing is changing–and for the better. In a world where public trust in science is being increasingly eroded, this book is not just for scientists at every stage of their career but for anyone who values research and the dissemination of knowledge."

Catriona MacCallum, Director of Open Science, Hindawi Ltd., and former Senior Editor of *PLoS Biology*

"Michael Hochberg is among the pre-eminent experts of scholarly publishing. His wisdom distilled into 'An Editor's Guide to Writing and Publishing Science' will be invaluable for young scholars developing their science writing skills as well as for seasoned authors seeking to put a bounce into their next publications. This is a book for us all."

Donald Strong, Professor, University of California, Davis, and Chief Editor of *Ecology*

An Editor's Guide to Writing and Publishing Science

MICHAEL HOCHBERG

Illustrations by
ALEX CAGAN

Great Clarendon Street, Oxford, OX2 6DP,
United Kingdom

Oxford University Press is a department of the University of Oxford.
It furthers the University's objective of excellence in research, scholarship,
and education by publishing worldwide. Oxford is a registered trade mark of
Oxford University Press in the UK and in certain other countries

© Michael Hochberg 2019

The moral rights of the author have been asserted

First Edition published in 2019

All rights reserved. No part of this publication may be reproduced, stored in
a retrieval system, or transmitted, in any form or by any means, without the
prior permission in writing of Oxford University Press, or as expressly permitted
by law, by licence or under terms agreed with the appropriate reprographics
rights organization. Enquiries concerning reproduction outside the scope of the
above should be sent to the Rights Department, Oxford University Press, at the
address above

You must not circulate this work in any other form
and you must impose this same condition on any acquirer

Published in the United States of America by Oxford University Press
198 Madison Avenue, New York, NY 10016, United States of America

British Library Cataloguing in Publication Data

Data available

Library of Congress Control Number: 2019937661

ISBN 978–0–19–880478–9 (hbk.)
ISBN 978–0–19–880479–6 (pbk.)

DOI: 10.1093/oso/9780198804789.001.0001

Links to third party websites are provided by Oxford in good faith and
for information only. Oxford disclaims any responsibility for the materials
contained in any third party website referenced in this work.

To my family

Preface

I could have written this book in 2009. I had just stepped down as chief editor of *Ecology Letters*, after 11 years of editing, spending several hours at it every day, and doing time on weekends. *Ecology Letters—ELE—*was my baby and, like any good parent, I wanted my child to grow and thrive. And that it did.

Starting in 1998 with 0 manuscripts, we gradually developed a following from ecologists dissatisfied with the usual delays in publishing. It took a few years to achieve our goal of 100 percent on-time publication decisions. Between 2001 and 2009, every single primary research paper submitted to *Ecology Letters* had a decision within 6 weeks. This promise branded the journal: "rapid publication of the most novel research." Decisions in under 3 months were uncommon in the 1990s and our innovative approach led to more submissions, a greater ability to select for quality and novelty, a higher impact factor, more submissions, etc. We built a reputation.

Our success was all about motivating people to achieve objectives. As chief editor, I interacted regularly with authors, reviewers, editors and, of course, editorial office staff, but also with publishers and production teams. I considered it essential to learn from both successes and failures. The secret ingredient was to follow each and every manuscript—*every day*—through the assessment process. So simple! Using an adaptive protocol, we achieved the milestones and forestalled the problems.

We received plentiful kudos—often from unexpected places. Despite our high rejection rates (90 percent), we very rarely had complaints—indeed, many authors of rejected manuscripts actually *thanked us* for having evaluated their paper rapidly. Publishing in *Ecology Letters* was a career-maker for many young researchers, who—before our journal—had no disciplinary high-profile alternatives to major journals like *Science*, *Nature* and *PNAS*. Other journals gradually viewed us as a model and improved their own services. When giving scientific seminars, I was always introduced as "the founding editor of *Ecology Letters*."

I could have continued my service, but in 2008, after 11 years and having handled over 8000 manuscripts, I felt that had done my job and was ready to move on. And it's a good thing I did.

I was able to maintain my scientific career during my tenure at *Ecology Letters*, and am indebted to many amazing collaborators. But now I was free from the daily responsibility of journal editing and this gave me an extra 20 hours a week to revitalize my research. Life with *Ecology Letters* gradually faded away, the only link being a yearly short course I gave on writing and publishing for PhD students at several European universities. Over the years of teaching, I saw how ill-prepared postgraduates were for scientific writing, and how perfectly clueless they were about the world of publishing.

I had to carefully think back to realize that I was just like them at the same career stage. In seeing the brightest students coming to my course largely helpless and leaving significantly able, I wondered if I could do more and for more people.

I decided to write this book.

About This Book

This book is a series of stand-alone chapters that take anywhere from 3 to 30 minutes to read. There are plentiful bridges between them. The reader is taken all the way from what she needs to know before even contemplating writing, to different phases of writing and its perfection. We then enter the world of journals and publishing before punctuating this with the submission, what to expect and how to react. Finally, I present several major challenges in publishing and opportunities in developing careers.

I wear three hats in writing this book.

I am first and foremost a scientist. I have written many papers with lots of people and for many journals. I ultimately learned to write science on the job, gradually honed my technique and—believe it or not—I still learn something new every time I write a paper. I don't consider myself a great writer, but I am a good writer—I'm a *professional* writer. Importantly, my experience as both a scientist and an editor has given me a unique approach to teaching writing. Unique does not mean untested: there are certain inescapable foundations to scientific writing that are found in any teaching—and they *are* here in this book. But in learning to write science, I also want you, the reader, to understand *what* you are doing and *why*. This book equips you with what you need to know about writing—and it will all make sense.

I am also an editor and know a lot of things about scientific publishing. These things live in the world of journals: the craft of committed editors and editorial staff, brilliant scientists who try to convince you to publish their work and expert reviewers who dedicate their time to assessing manuscripts. There are also difficult and sometimes unpleasant things: tough publication decisions, incomplete information and delays, conscious or unconscious biases in manuscripts and reviews, and disappointed or even outraged authors. Few scientists really know about how journals work and the world of publishing. This book opens these black boxes and will help you write better papers and publish them more intelligently.

Finally, I've been a part of the scientific community for more than 30 years and have witnessed the rise of personal computers, the advent of the Internet and the growth in the number of practicing scientists and of scientific journals. More powerful, faster and bigger has catapulted science into the twenty-first century, but not without new challenges. These include our abilities to assess and improve science, gain access to published material, cover publication costs, and decide what science is worth reading and citing. These challenges are being met in the world of "Open Science," a community of initiatives that will make all aspects of science more transparent and as accessible as possible. Open Science is moving *very* fast and I only provide a snapshot of this burgeoning revolution. I highlight how you can flourish both as a scientist and member of the scientific commons in this emerging landscape.

My Style

This book is written for researchers in the biological sciences, though much of the material will also be applicable to the physical and social sciences. The scholarly mantra—accuracy, neutrality, clarity and precision—applies to all of us. Interdisciplinary likenesses also extend to journals and publishing, including criteria for choosing a journal, replying to editors and reviewers, and broader issues such as the sustainability of different publishing models. Nevertheless, the devil is in the detail, and this book does not navigate all of the interdisciplinary contrasts. Readers from areas other than biology should be aware of this and use the information in this book accordingly.

Exercises are essential for improving writing skills, and key chapters will challenge the reader to both do and learn. However, this *is not* an exhaustive, step-by-step book of exercises and answers. Exercises are only part of the writing equation, and what is largely missing from other books on writing—and provided here—is the greater context. This book is written to make the reader stop and think. Some of the tools I use are personal experiences, behind-the-scenes observations, opinions, boxed highlights and take-home messages. A handful of these messages are "Golden rules"—this book's central insights. The Golden rules are highlighted in Alex Cagan's wonderful illustrations.

This is a book of deduction, perspective, opinion and suggestion. The worlds of writing and publishing have few hard-and-fast rules. There is no single *bone fide* approach to writing a scientific article and no "United Nations of Journals" to regulate author, reviewer, editor and publisher behavior. The practical consequence is that I have had to repeatedly commit one of the cardinal sins of scientific writing: weasel words. This book is *full* of "many," "some say," "others view," etc. With this in mind, I ask for the reader's indulgence. Just like reading scientific articles themselves, you will need to form your own opinions, and do fact-finding and fact-checking. I do nevertheless provide key references and recommendations for further reading, so that the interested reader can dig deeper into the debates and the data.

Acknowledgments

I have interacted with many special people over my scientific and editorial careers. Thank you to Jan Volney for supervising my first project leading to a publication, and to Andrew Leibhold, David Wood and James Milstead for commenting on this first paper. To Michael Hassell and Jeff Waage for guiding me through my PhD, and to John Lawton the same for my postdoc. To my many collaborators and, in particular, Bob Holt and Joel Cohen for their influence on my writing. To Robert Barbault and Jean Clobert, who convinced me to come to Paris to do science. To Nicole Pasteur and Isabelle Olivieri, who enabled my move to Montpellier, and to Jean-Christophe Auffray and Agnès Mignot, who have continued to give me the precious freedom to do research at the Institut des Sciences de l'Evolution, University of Montpellier. I am indebted to the CNRS and the James S. McDonnell Foundation for their research support during the writing of this book.

My editorial career and this book would not have been possible had it not been for three events. The first was Robert Barbault asking me to be chief editor of *Acta Oecologica* in 1996, which ultimately led to my becoming chief editor of *Ecology Letters* from 1998 to 2009. It was during my tenure at *Acta* that I discovered that it should never take more than 2 months to make a publication decision. The second was a meeting with Blackwell Science that led to the founding of *Ecology Letters*, and I thank Bob Campbell, Aileen Boyd Squires, Robert Barbault and Simon Rallison for their inspiration in launching the journal, and to Liz Ferguson, Lynne Miller, Debbie Wright, Nathalie Ferrand, Françoise Gaill, Marie-Louise Cariou, Bernard Delay, Pierre-Henri Gouyon, Stéphanie Thiébault, Martine Hossaert, Francine Roussel, Christelle Blee, Anne-Sybille Loiseau and Nathalie Espuno for their support in continuing the adventure. Special thanks to my editor-in-chief successors—Marcel Holyoak and Tim Coulson—for keeping the flame alive. The third event was being invited to teach a course on writing and publishing scientific articles at the University of Helsinki in 2006, which led to teaching the course at numerous universities in Europe. I thank Hanna Kokko and Anna-Liisa Laine for the original invite and the many other organizers who hosted the course, which forms the backbone of the present book.

True to my recommendations at the end of Chapter 5, I have written this book in several places. I would like to thank Ingela Alger, Paul Seabright and Valérie Nowaczyk at the Institute for Advanced Study in Toulouse; Jennifer Dunne, David Krakauer and Tom Real at the Santa Fe Institute; and Oliver Kaltz, Claire Barbera and Emanuel Fronhofer at the Institut des Sciences de l'Evolution, Montpellier. In the years leading up to this book I also benefited from comments and discussions with Ana Rodrigues, Dries Bonte, Hildegard Uecker, Alex Roulin, Slimane Dridi, Jorge Peña, Adin Ross-Gillespie, Astrid Hopfensitz, Charles Fox, Denis Bourguet, Daniel Schrag, Carl Bergstrom and Vincent Calcagno.

I would like to extend sincere thanks to those who made this book possible. Discussions with Joshua Schimel, John Miller, Peter Turchin and Geoffrey West were invaluable in understanding what it takes to climb this highest of mountains. A special thanks to Don Strong, Catriona MacCallum, Liz Ferguson and Art Weis for taking the time to comment on, and greatly improve, the manuscript. To Alex Cagan, for bringing this book to life with his marvelous illustrations (the featured character depicts me at different points in my career!). To Oxford University Press and particularly Ian Sherman and Bethany Kershaw for their tireless support. Finally, to my wife Joëlle for giving me the sometimes-unreasonable space I needed to write.

Contents

I. BEFORE YOU BEGIN

1. Planting Your Flag — 3
2. Quality and Productivity — 6
3. Citing, Reading and Searching — 13
4. Avoiding Plagiarism — 22

II. WRITING A GREAT PAPER

5. The Writing Mind-Set — 27
6. The Start — 32
7. Use *Models!* — 42
8. IMRaD — 50
9. The Vitrine — 60
10. The Puzzle — 68
11. Emphasis and Finesse — 72

III. CHOOSING WHERE TO PUBLISH

12. How Journals Operate — 81
13. Who Really Decides? — 91
14. What to Expect from Journal Service — 97
15. Choices in Publishing — 104
16. Choosing a Journal — 114

IV. SUBMISSION AND DECISION

17. Authorship — 125
18. The Cover Letter — 133
19. The Publication Decision — 139
20. Data Archiving and Sharing — 146

V. CHALLENGES

21. Peer Review	153
22. The Cost of Publishing	164
23. Use of Citation Metrics	172
24. Disposable Science	183

VI. OPPORTUNITIES

25. Developing Your Career	191
26. Collaborating	199
27. Writing Reviews, Opinions and Commentaries	205
28. Reviewing Manuscripts	209
29. Social Media	218
30. Old Dogs, New Tricks	223
Glossary	229
Notes and References	235
Suggested Reading	249
Index	251

PART I

BEFORE YOU BEGIN

Science needs to be communicated to exist and to evolve. Scientific communication can take many forms, but it is the published article that is the immutable tablet that a community of thinkers can read, debate, cite and build upon. Publication is the scientific community's dialogue. It is knowledge. Sure, the scientific edifice can be fragile, but that does not matter. Right or wrong, we learn and gain insight. We communicate it and science advances.

Science is a craft. We apply clear and repeatable methods to test hypotheses and uncover clues. We communicate our findings by writing accurately, neutrally, clearly and precisely. We are self-critical: we replicate, express caution and re-test. Science is based on facts, on evidence, but we nevertheless tell a captivating story—the beauty of the quest and its findings—the ingenious and rigorous application of the craft.

We do science because:

We are curious and want to know how the world works.

We enjoy resolving puzzles and problems.

We get recognition and advance our careers.

We want to influence others and the world.

But a scientist's freedom can also be science's worst enemy. When young researchers join the community they learn the craft, but may feel under pressure to emphasize career motives. Careerism does not necessarily lessen responsibility, but the former—if pushed too far—can erode the latter. More low-quality, biased publications, lack of reproducibility and scientific misconduct leading to retractions all do a disservice to science.

This book is about how to write effectively and publish intelligently, and in doing so responsibly, meet your scientific and career objectives.

This section introduces the reader to issues in responsible science. The first is simply that science only exists when it enters the community—that is, it is published (Chapter 1). Developing a research program and career means considering the importance of—and balance between—quality and productivity (Chapter 2). The quality of the craft enters into writing the manuscript, particularly scholarship (Chapter 3), and avoiding plagiarism (Chapter 4).

1
Planting Your Flag

Science works through the publication of results. Sometimes results confirm previous findings, others they are a new discovery. Either way, results only have effects when they are actually made available to the scientific community. Delaying publication may mean that what was a discovery is now a confirmatory finding. This chapter explains why it is important to "plant your flag" as a personal achievement, a necessity for career development and a gain for the scientific community.

Discovery is one of the hallmarks of science. A new finding advances knowledge and creates more questions for future inquiry. There is a myriad of reasons why we value discovery, including satisfying curiosity and achieving advancement. Some are motivated by recognition, prizes, invited seminars or by directing large successful research groups. Others are on a never-ending quest for the truth. Regardless of the blend of motivations, discovery is a primary driver of science and scientists.

Discovery is a new and interesting finding—it's novel. A novel finding solves an outstanding problem and opens new lines of research. But novelty is a subjective concept. What is incredible and highly relevant to one may be mundane and tangential to another. This observation is important, since it suggests that journal editors, peer reviewers, scientists at large, the media and the public—five kinds of readers that your results ought to influence—may each have different views on the significance of a finding. Their views will be influenced of course by their own education and experience, but also by the way findings are presented. This puts a premium on how a manuscript is written—the main theme in Part II of this book.

Although a major discovery is the dream of every scientist, most science does not push the frontiers: it is incremental or confirmatory. An incremental study makes a novel advance, but does not fundamentally change the way scientists view a problem. A confirmatory study on the other hand seeks to determine whether a previous result holds when tested using the same or under new conditions. Such a study may or may not produce new insights, but its importance lies in evaluating whether current understanding is correct, limited or wrong. Just like discovery, viewing a study as incremental or confirmatory is a subjective assessment.

Regardless, new results—whether novel or confirmatory—have a place in science. But to have this place, they need to be accessible and referenceable by the scientific community.

Science must be retrievable, read and citable by scientists.

Science must be published.

Recognizing and Crediting Discovery

A key concept in science is that a given discovery can only be made once—it has scientific priority over subsequent new discoveries, increments and confirmations.

> *Henry Oldenburg* launched the first science journal—*Philosophical Transactions*—in 1665,[1] largely in response to concerns that credit for discovery was either verbal or through correspondence and had no central third party authority. Prior to *Phil Trans*, discoveries were publicly acknowledged (as witnesses) before findings were shared with the science community. *Phil Trans* established itself as an independent third party that published discoveries and their dates of submission.

Priority emerges as a multiple step process.[2] In Henry Oldenburg's day, this began with "disclosure," and some (but not all) scientists today do announce findings either at a scientific meeting or, increasingly, in the form of a preprint (see Chapter 15). The second step is publication in a peer-reviewed journal, although the posting of a preprint effectively makes the priority of an official publication redundant. The final step is recognition of priority through citation. This is the most subjective and error-prone of the three steps. Crediting a discovery requires that a scientist finds the paper in the first place *and* can assess its relevance, which means a good grasp of the literature. Finding and integrating the literature are particularly challenging. The number of articles published annually continues to grow and past publications are forever viable. New disciplines emerge, and existing ones are increasingly sub-divided into ever-more specialized and technical areas. More volume and greater diversity make attribution of an original finding hard to assess. The end result is that crediting discovery is often open to interpretation, based in part on one's own (limited) knowledge of the literature.

Discovery and its priority therefore are not absolute. Often, different schools of thought will have different views, and the recognition of who did what and when, and why it was important, will change over time. For this reason it's impossible to evaluate the significance of a finding at publication. A result may appear to be a breakthrough, but with time turn out to be either wrong or less novel than originally thought. A finding may at first appear mundane only later—once science is ready—to be found as remarkable.

Plant Your Flag

We are sometimes not sure whether our results are ready or interesting enough for publication. "There just isn't enough here for a complete study," or "With a little more work, I might discover something big!". Indeed, continuing work may be the best option, but ultimately the time *will* come when the decision to publish has to be made.

Scientists learn to recognize this and *why* this decision is so important.

A discovery is only a discovery once published. Waiting too long can mean that the same discovery is published by someone else first, which means that your finding is either relegated to the position of "second demonstration," or, if very delayed, possibly viewed as a confirmation. Publishing too early, however, can mean that the scientific quality of your discovery is limited (see next chapter). Rushed, low-quality studies can mar one's scientific reputation. Advice from senior colleagues can be invaluable here to evaluate the readiness of a study for publication.

Publishing is metaphorically like an explorer climbing a mountain and planting her flag at the summit. Planting *your* flag:

- is a claim to priority;
- is a gain for science and the scientific community;
- maintains research momentum;
- is a personal achievement;
- is important for self-esteem and gaining the esteem of colleagues;
- makes a reputation;
- is central for career prospects.

2
Quality and Productivity

Authors need to write productively while maintaining quality standards. Productivity pushed too far can negatively influence quality, which can mean publication in less demanding journals and a lowering of one's scientific reputation. This chapter discusses the essentials of quality and productivity.

"Publish or perish" is probably the most hackneyed axiom in science. To be sure, and as emphasized in the first chapter, we publish to scientifically exist—and indeed a career in science *requires* a satisfactory rhythm of publication. For some scientists however, publication is more than just the culmination of a study: the more productive I am, the more funding I get, the more I grow and the more science I do.

Emphasizing productivity can be a good thing. Productivity largely goes hand-in-hand with synergism: the science is better, otherwise-unattainable projects become reality and scientific careers are made. But if pushed *too* far, productivity risks sacrificing scientific and writing quality. The reason is simple: each person and each research group has limits on how much they can accomplish in a given time frame, before depth, clarity, accuracy and precision suffer.

To clarify the issues relating to the limits of quality and productivity, just for the sake of argument, consider the extremes of exclusively valuing either one or the other.

Quality

Quality is the level of a finished product relative to a standard reference. This includes scientific parameters, such as scholarliness and reproducibility, and communication parameters such as clarity and accuracy in writing. A "standard" is the specification of what is expected, and is characterized by the completeness, lack of bias and precision of the content. Thus, for example, a statistical analysis of a high standard will be better at fulfilling assumptions than a lower standard analysis and will be correctly executed.

> *The mantra of rigor—accuracy, neutrality, clarity and precision.* Scientific writing is the accurate portrayal and interpretation of your observations. I transcribe what I see. I'm neutral. If I am writing an *Opinion*, *Commentary* or *Perspective*, then it is *expected* that I present viewpoints or speculate (see Chapter 27). But if I'm casting an original research article, then although inferences and opinions are certainly allowed, they should be used sparingly and be signaled—"This result suggests…" or "We speculate…". Accuracy and neutrality are not the only pillars of rigor. Writing must be clear and precise. Accuracy is similar to clarity and precision. Whereas accuracy refers to being factual, clarity means that the *language* used unambiguously transmits what the writer intends to say to the reader. Precision, on the other hand, is exactness. Precise writing zeros-in on a concept so that the reader appreciates its significance.

Quality takes time. Time to read, think, integrate ideas and formulate solutions to the many questions and problems encountered in conducting and writing a study. Time also lets the writer look afresh at her manuscript, correct errors and ambiguities, and increase clarity.

Quality requires investment. As quality is pushed higher, we tend to complete fewer, but better studies. There are practical reasons for seeking high standards, including increasing the likelihood of publication in a prestigious journal and the impact[1] of the paper once published. Authors may also have philosophical reasons to seek quality, viewing high standards as necessary to leave a useful and durable imprint on science. Quality is both a reputation builder and a service to the scientific community.

> *Quality is not necessarily impact.* Quality—although respected by scientists and more useful to science—does not necessarily translate into impact, and hugely influential articles can be poorly written and of questionable scientific quality. This is because impact is influenced by so many other factors, including timeliness, interest, journal of publication, author identities, coverage in seminars, buzz, etc. For more on impact, see Chapters 23 and 24.

Investing in quality beyond a point, however, can waste time better spent otherwise, or even be detrimental. There are several interrelated reasons for this.

First, we may believe that doing more increases scientific standard, for example, the number of experimental replicates. However, beyond a point, adding more does not improve the study. Similarly, investing in quality writing is useful to a point, but past this, dotting i's and crossing t's is a waste of time, and time wasted means lower scientific output.

Second, taking too much time to improve quality may mean that you get scooped by another research group who happened to publish their findings (and receive credit for priority) before you.

Third, conducting research in the (very) long term without publication carries the risk of losing interest in the topic. People change. Also, research areas change. What you and the scientific community view as exciting today may be passé in several years' time.

Finally, we live in a world where the time required to execute high-level science can be at odds with impatient collaborators, students and postdocs who need to publish, and with granting agencies that expect you to publish in a reasonably timely way. Publishing years after what was originally planned risks irking collaborators, penalizing students and postdocs, and weakening grant applications.

Productivity

Productivity is a far easier concept to grasp compared with quality. It is the number of papers published in a time frame.

Intuitively, more papers necessarily mean less quality in each paper—and, as mentioned above, beyond a point this is true. I couldn't possibly publish 20 papers a year and maintain quality levels that would be acceptable in high-standard journals. Impossible for me, but perhaps there are those who could. This underscores an important observation. Although *all* scientists (and research groups) are subject to the productivity–quality curve depicted in Figure 2.1, the onset and extent of the trade-off

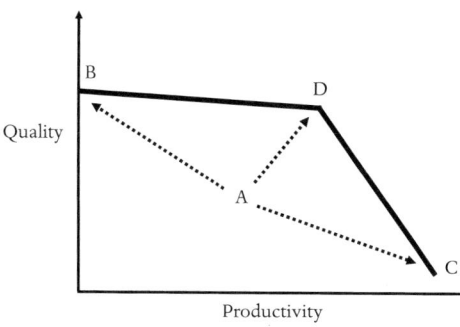

Figure 2.1 Hypothetical set of possible productivity–quality strategies (solid line and area below it) for a given researcher. If the scientist is currently at point A, she maximizes quality by reducing productivity and attaining B, or increases productivity at the expense of quality to C. She could also seek an combination (e.g. D) that increases both quality and productivity.

will differ from person to person (and group to group). So, my productivity–quality curve may have the same basic shape as yours, but be shifted upwards and to the right.

Productivity seeks numbers—more articles published. It can be accomplished by working more hours, doing easy-to-conduct science or aiming for minimally publishable units (MPUs or so-called "salami slicing"; see below).

Productivity pushed too far has several significant drawbacks.

First and foremost, beyond a certain point, more papers per unit time increases the likelihood of inaccuracy and errors. Aptitudes such as background knowledge and writing skills can make a major difference in how much someone can accomplish without appreciably sacrificing accuracy. But everyone has their limits. Ambiguities, factual mistakes and more minor misapprehensions in citations are picked up by colleagues. Errors in analysis can sometimes be catastrophic, possibly resulting in an article being retracted.

Second, reputations are hard to gain but easy to lose. Attaining productivity while maintaining quality takes time and experience and, as readers encounter your name, they associate you with the quality and interest of your work. Publishing few high-quality articles will draw greater respect than churning out lower-quality papers. Low quality and errors can permanently taint a scientific reputation.

Third, pushing productivity at the expense of quality will likely relegate your manuscripts to publication in journals not recognized by authorities such as Web of Science (WoS), Scopus or the Directory of Open Access Journals (DOAJ)—or you may be tempted to publish in a "predatory" journal (see Chapter 15). Because peer review in these venues is of lower standard, published articles will retain errors, imprecisions and unsubstantiated claims that otherwise would have been corrected or removed. Moreover, your work will be largely unseen by readers who ignore papers in those journals.

The publication of large numbers of low-quality articles lowers standards, confuses debates and slows scientific growth.

The reproducibility crisis. Career incentives to publish more, faster and in higher-impact journals has led to a situation[2] where (i) to be accepted, papers usually need to have positive results; (ii) experiments are not always replicated; and (iii) research practices are sometimes questionable. These factors contribute to publication bias in the established literature.

Quality and Productivity

Exceptions do exist, but scientists who strive for quality to the point that they very rarely publish, and those who saturate the literature with low-quality work are unlikely to have influential careers. Most are aware of the pitfalls of each extreme.

Scientists have considerable latitude to develop their productivity–quality strategy as portrayed in Figure 2.1. There is no single optimal strategy. Some people may feel comfortable at point A, or will develop their potential in productivity and quality, respectively, to move from extremes at B and C, to the compromise at strategy D. The reality however is that a scientist functions in a world where impact influences careers. To the extent that overall impact increases with productivity, there is the danger that successful strategies are not at point D in Figure 2.1, but rather at C. Although positive for the individual scientist or group, the overall effect on science can be negative.[3]

Personal goals and abilities will influence where you are on the productivity–quality curve, but often you will be collaborating with other scientists and *they too* will have their own productivity–quality relationships. This can make collaborative situations challenging for the simple reason that the curves of the various participants may be substantially different from one another. This may be because different participants attach different weights to quality versus productivity, or because some participants are more able to work fast while maintaining high standards compared with others.

When is it One or Two Papers?

At the heart of the productivity–quality issue is "When is it a paper?" and "When is it more than one paper?". The simple answers are, respectively, "When a study makes a new contribution" and "If there are two or more distinct contributions that each stand on their own." Journals offer little additional advice in their guide to authors, usually just referring to scope, article types, and word, figure and reference limits.

> *MPUs as a strategy.* Research programs and career advancement evidently depend on publication, but the average time necessary for a graduate student in the life sciences to publish their first paper has increased relative to three decades ago, and is approaching the typical duration of a PhD.[4] This is likely due to the requirement for more data for publication, but also increased gatekeeping at reputable journals. One solution to this quality–rapidity dilemma is aiming for MPUs, or a mixed strategy of MPUs (submitted before the dissertation is completed) and more consequential original research papers (submitted either before or after the dissertation).

The most effective way to decide whether your study is a "paper" is to read articles in the target journal. Should the depth and breadth of your study sit comfortably within the range of these articles, then you will have some indication that a submitted manuscript will be seriously considered. Making this judgment can be subjective, and so it is a good idea to also ask the views of a colleague.

You may however encounter the situation where the MPUs at journals of interest indicate that your study *could* produce more than one viable manuscript. Ideally, you would have made this assessment when your study began, but science is never completely predictable!

In deciding whether your study is one or more manuscripts, the following may prove useful.

1. A published study generally develops a central result. If you believe that you have at least two central results, then you may have reason to write two (or perhaps more) manuscripts. However, writing two manuscripts entails more work than writing just one, and there is no assurance that both will be published. Even if both are published, a weak paper may be largely ignored, raising the question of whether a combined paper would have had more impact.

2. Are you going to first decide on one or more manuscripts, write them up and *then* explore journals, or the reverse? Doing the former may constrain journal choice, whereas choosing the latter may affect what part(s) of your study to include, with the risk that some results are tangential and insufficient to stand on their own in any journal.

3. You may find that some results simply do not contribute to a cohesive story. The natural reaction is to either orient the write-up to accommodate the results, or to see how to create an additional manuscript around the results in question. It may be that neither solution is adequate and that these results are never published.

The evaluation culture is the dependence of scientific careers on how much we publish and where we publish. Productive groups that publish in top-ranked journals receive more funding, more and better students, more awards, etc.—they gain academic prestige. Governments, institutions and funding bodies that promulgate these measures of "impact" in their decisions, foster career science. Several initiatives are addressing the issues surrounding this phenomenon, notably the Declaration on Research Assessment (DORA).[5]

Advice

The evaluation culture pushes scientists to publish more and in top-ranked journals. The quandary for young researchers is that they learn to be masters of the crafts of quality science and writing, but also need to publish reasonably quickly, publish a lot and in reputable, top-ranked journals.

How can you attain your career goals through healthy productivity, but without sacrificing quality?

1. *Discuss with more experienced scientists.* It is ultimately up to you to decide the emphasis on quality in your own research and writing, knowing that certain standards will be necessary in for publication in reputable (even if not the highest ranked) journals. The development of your perspective will benefit from discussions with mentors, principal investigators (PIs) and more senior colleagues.
2. *Publishing relatively short manuscripts need not be "salami slicing."* Ask yourself whether your study has attained the breadth and depth that you seek. Will an extra segment (e.g. experiment, analysis) be value added? Time and logistics need to be factored into this assessment.
3. *No single size fits all.* As you go through your career, you will encounter other researchers on the full spectrum of productivity–quality strategies. It is important not to sacrifice the quality part of your *modus operandi* simply because certain colleagues are highly productive. With experience you will increasingly achieve greater productivity without sacrificing quality.
4. *Observe models.* Models can be supervisors or mentors, but also scientists whom you do not know personally, but through discussion or studying their publications you learn or infer publication strategy. Observing models and thinking about what you admire (or do not) in their publications will be useful in developing your own approach.

3
Citing, Reading and Searching

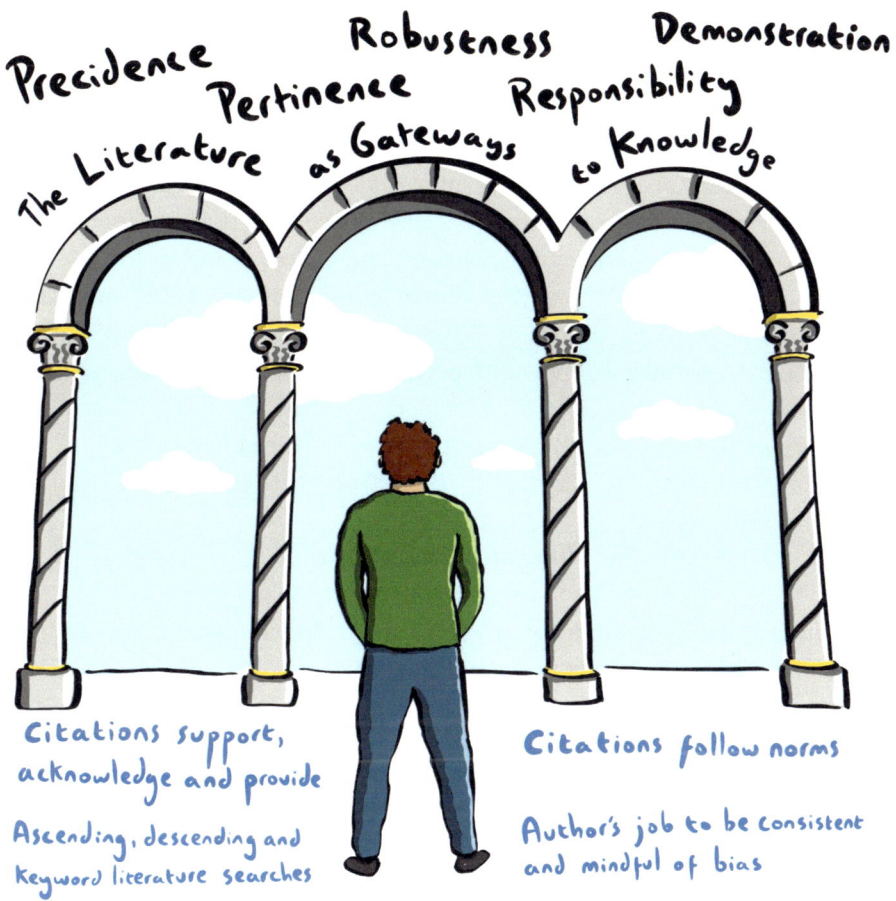

Citation supports claims and provides readers with links to dedicated study. With an ever-growing literature however, consistently citing relevant work is becoming increasingly difficult. In this chapter I discuss citation, biases and good practice. I also stress the importance of reading the literature in building knowledge, and present methods for effective reading and literature searches.

Science builds on what has come before. The record of this appears in the form of citations. Citations provide evidence for our statements and claims and recount the history of intellectual growth. But the corpus of published work is ever-growing, increasingly diverse and specialized. This puts authors at increasing pains to accurately choose what to read and what to cite from the growing pool of relevant literature.

There are no hard and fast rules for citation. We often cite based on what we learn from our mentors, in interacting with colleagues and—perhaps mostly—what we infer from published articles themselves. Authors are largely left to their own devices in how they cite. Different people cite differently because they have different perspectives on or knowledge about what is important to cite. Moreover, sometimes they do not fully understand the functions of citation, or do, but have no consistent method.

Citation is a scientific act, similar to experimental design, statistical analysis and interpreting results. Citation is however more subjective than these and other aspects of science, and it is for this reason that many researchers dedicate little time and effort to scholarly citation. Because journal reviewers and editors recognize this and themselves are time-constrained, they often simply ignore citations in their assessments. This gives authors a degree of freedom in deciding what to cite and opens citation to possible biases. In learning the craft of science, a young researcher therefore needs guidance into the world of citation.

This chapter discusses why we cite, and practices and issues in citation.

Reasons for Citation

A citation is the declaration that a previously published article is relevant to the present study. There are three different bases for citation.

1. *Evidence supporting claims.* There are numerous ways to present information. This can range from statements of fact to suggestions, and from the obvious to the surprising. Depending on where a statement lies in this space you will need to decide whether or not it should be supported by citation. As you go toward "fact" and "surprising" you probably need to cite. You need not cite a totally obvious fact, but you should consider citation support for a not-so-obvious suggestion.

2. *Acknowledging discovery, results and ideas.* This is subtly different from supporting claims. A claim is potentially controversial: it is believed or not believed; known or not known. In contrast, a discovery, result or idea has a definite source. So, if you start a sentence with "Previous work showed . . .", then you obviously need to include one or more citations.

3. *Gateway to information.* Citations are necessary when making a claim or acknowledging a previous finding. But citations are also a gateway to information. One of the roles in being an author is—as an expert in the field—to inform the reader. Sometimes this is done by citing original research articles, but more often links to information are made by citing surveys, such as review or synthesis articles.

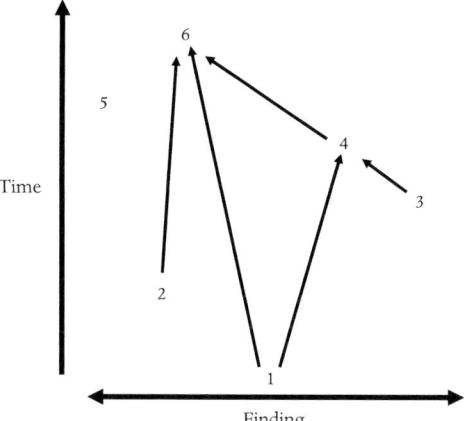

Figure 3.1 *Tree of discovery.* Hypothetical citation tree relating a discovery (study 1) to five subsequent publications (in chronological order). The discovery and subsequent refinements are represented on a simplified "finding" axis (e.g. the finding in study 3 is maximally different from study 5). Thus, studies 4 and 6 cite the precedence of study 1, whereas 2, 3 and 5 were either unaware of 1, or had other unknowable reasons for not citing it. The most recent study, 6, cites 1, 2 and 4, and may have seen study 3 cited in 4, but did not cite 3, possibly because of insufficient relevance in the findings. Finally, study 5, which produced a finding similar to, for example, 1 and 2, did not cite these, and study 6 did not cite 5 despite the thematic proximity of the findings. It is possible that studies similar to 5 and 6 and published after the latter will cite both, thereby consolidating the tree's history.

Citation Criteria

Like many other aspects of scientific writing, citing previous work is usually learned on the job. We infer or learn it from the articles we read, from websites, PIs or from colleagues when actually writing a paper. Given the diversity of sources and infrequent correction—particularly following peer review at journals—it is not surprising that individual scientists vary considerably in the criteria they employ and how consistently each is applied.

Read (much) more than cite. According to one source[1] the average university academic reads about 260 articles a year and spends an average of 49 minutes per article. Expect to read far more articles than you cite. Some non-cited papers provide background, others are directly pertinent but not priority citations and yet others turn out to be tangential. In my subject area (population biology), for a typical original research article with 50 articles cited, I will have likely read more than 100 articles.

The main criteria for citation are:

- *First demonstration.* This gives credit to discovery and enables the reader to interpolate the path of knowledge from discovery to the present. As discussed in Chapter 1, it is not always straightforward to determine "discovery" and sometimes early, robust demonstrations are more justifiably cited compared with an earlier undeveloped idea.
- *Recent demonstrations.* Science can move very fast—flags are planted all the time. Because science builds on growing and changing foundations, your study will probably be more relevant to recent work than to studies from the distant past. Citing recent work conveys information about the context into which your own work enters. However, some authors are wary of citing *very* recent publications until they are recognized (cited) for their importance, either by experts in the field or through consensus (see below for possible bias in the latter).
- *Diversity in demonstration.* Scientific knowledge grows and evolves. A given question may have more than one answer. Citing and discussing alternative views or conflicting results provides a more balanced appraisal than only citing the majority view or the author's own school of thought.
- *Robust demonstrations.* Ideas are everywhere and many published ideas are one-liners or exposés that lack in-depth development. Even when an analysis is conducted, rigor and depth can vary considerably from study to study. All else being equal, citing carefully developed ideas and investigations does a greater service to science than work that is somehow limited or does not make a significant advance.

Issues: Arbitrariness, Subjectivity and Citation Bias

Who hasn't read an article and thought: "Why did they cite *this* paper?" or "How could they *have possibly missed* citing that paper?"

Citation is far from an exact science. Just a few of the many culprits:

- editor and reviewer apathy and tolerance;
- author non-mastery of the literature;
- citation bias.

These three points argue for editor and reviewer responsibility, scholarship and balance, respectively. The most damaging of these issues is however citation bias.[2] Citation bias occurs when authors cite from (or avoid citing from) a particular subset of articles. Citation bias may be conscious and even strategic, or unintentional and inadvertent.

> *Distortions.* Greenberg[3] investigated a complete citation network of papers addressing the belief that "β amyloid, a protein accumulated in the brain in Alzheimer's disease, is produced by and injures skeletal muscle of patients with inclusion body myositis." The author analyzed how certain publications sometimes created distortions in the original result and how they were perpetuated by subsequent citing articles. These included: bias against critical primary data, bias to justify models, invention and altering the meaning of the result. The perpetuation of distorted claims created a snowball effect of gained (unfounded) authority in the source of the distortion and further perpetuation of the claim.

Many citation biases stem from article characteristics or author identity (also see discussion in Chapter 24):

- journal in which the article appears;
- statistical significance of results;
- number of times the article has been previously cited;
- country or ethnic origin;
- gender;
- author impact;
- author career level;
- self-citations, citing colleagues and friends.

Regardless of its cause, citation bias does a disservice, since less deserving papers/authors are given credit and the more meritorious are ignored. Discerning, expert readers will form their own opinions about what *they* would have cited or not cited, but evidently cannot change what they read. Worse, undiscerning non-experts risk accepting what they read and then perpetuating biases when citing in their own papers.

Responsibility

Authors have a duty to inform readers of the relevant literature, but there are constraints involved. First and foremost, journals usually impose maximum numbers of citations allowed. Authors need to establish "necessary" citations, and then choose which of the more facultative citations to maintain. To accomplish these decisions in a balanced way, it is best to cite as extensively as possible and—once the manuscript is complete—make the difficult choices as to which references are unnecessary and can be removed.

Reviewers and editors have important roles in arbitrating citations, but clearly because of the Herculean work involved in verification, cannot be as invested as the authors themselves. Reviewers should consider it their responsibility to be mindful of faulty citations and suggest missing relevant ones. Some editors will reiterate or comment on the reviewer's reference suggestions in their own comments, but many editors ignore these issues and leave it to the authors to decide whether or not to heed a reviewer's suggestions. This can have the unfortunate consequence that authors cite whatever the reviewers suggest (including the reviewer's own articles!) for fear that if they don't, then their paper may be rejected.

Reading

Good scientific writers are also good readers. The reason is simple: much of the content that goes into a paper comes from published work. This includes basic principles, methods, unsolved questions and important findings. Reading provides both the foundation and scaffolding for your own scientific contribution. Reading is the basis of the papers you will cite.

We develop knowledge and understanding through reading in different ways. Some reading is not specific to a particular project or manuscript and, rather, develops curiosity and general culture. Other reading will happen during the preparatory and execution phases of a study, and this includes evaluating what has been done previously, and identifying outstanding questions, methods and logistics. Finally, you will augment any previous literature survey with reading during the writing of a manuscript, specifically to develop and support arguments and find evidence relating to your study's findings. As you write, you will likely encounter new, unexpected queries and problems requiring additional reading.

In writing your paper, you will have made assumptions about its likely readership. Readers could range from students to experienced experts within your area, to interested people from other disciplines. Your writing will need to have a consistent standard in expertise, but also go into depth in places key to developing the big picture. If you are to achieve this and guide the desired range of readers, then you will need to be at least as expert in your own subject as they are. By the time you finish writing your paper, you will have done *a lot* of reading!

Strategy

There is no limit to how much can be read on a subject. If you have a computer folder with 300 relevant articles, will you read them all completely? Possible, but unlikely. Imagine that you read 10 papers a day. Besides being exhausted, you will have spent a month doing little else other than reading. Worse, although reading educates, you will ultimately discover that many of the articles in your folder are only tangentially relevant to your manuscript. Could you have done this differently?

You need a reading strategy to efficiently master the relevant scientific literature. Let's revisit the above scenario. Now, rather than reading 10 articles every day, you decide to read each and every one *very* carefully. Two hours is a reasonable estimate to dissect an important article. So now the task takes 2 full months, and you are probably thinking about an even longer vacation! The point is that some of your 300 articles will only take 5–10 minutes (probably will not influence your paper)—others up to 1 hour (probably to be cited)—and just a few important ones beyond 1 hour (definitely cited and possibly to be discussed in your paper).

How can you put this into practice?

The goal is to decide in just a few minutes whether a paper is worth being read further.

Step 1 Look no further than the first page of an article—there, right in front of you is the lure set by the authors—the title—and should you be tempted by a condensed explanation of the paper—the abstract.

Step 2 Some skip this and go directly to Step 3, but I suggest that you take a couple of minutes to look at any figures. Figures are likely to display the most important results and insights.

Based on the title/abstract/figures, decide "yes," "no," or "maybe" to start actually reading the paper. If "maybe" make notes why. You can return to "maybe" articles later.

Step 3 If you decide to read, then start at your usual pace. Some readers will decide partway through that the paper is not worth reading. If you do continue, then adjust your reading speed according to relevance to your own work. Make notes on the document itself—if it is a pdf file, then you can electronically highlight text and insert notes. An experienced reader will spend the time necessary on important parts and gloss over others.

Organization is a crucial to effective reading. This includes the easy discovery of papers based on keywords or phrases, making notes and highlighting parts of a paper that you believe are important, and finally taking a few minutes after reading to note how the article may contribute to your own study.

Literature Search

Similar to reading itself, actually finding the literature can be a considerable undertaking. There are too many publications and too little time to sort through and read those that appear relevant. Nevertheless, surveying the literature is important, since it provides the raw material for reading, potentially influencing your work.

> *Problems and solutions.* There are numerous challenges to accessing and processing the literature. First, many readers have limited or no access to published material due to subscription paywalls. Paywalled articles can sometimes be obtained free of charge by contacting the corresponding author and requesting a pdf copy. Although potentially subject to legal pursuit by copyright holders, virtually all paywalled articles are available via SciHub.[4] A second issue is the cumbersome nature of literature searches (see below). Semantic Scholar[5] uses machine learning applied to keyword searches to find what is most relevant and impactful. A third issue is discovering citations themselves. Open citations[6] can identify networks of relationships between papers that facilitate discovery (see Figure 3.1). Achieving open citation will require the extraction and organization of these data into a centralized, discoverable database at Crossref.

Although literature surveys can never cover 100 percent of all publications in existence, there are sources and approaches that will enable you to consistently go deep into the literature.

Sources of potential articles of interest include:

- colleagues;
- seminars and conferences;
- social media;
- journals.

Use of these outlets, such as table of contents emails from journals, can provide a sample of articles, but more systematic searches require databases—either personal compilations or Internet sources.

Three techniques that can be employed with the aid of dedicated databases such as the Web of Science are:

- keyword search;
- ascending search;
- descending search.

You will surely be familiar with keyword search. The latter two methods are less well known, though we often use them without actually realizing it.

Keyword search

The use of keywords has evolved considerably since the advent of the Internet. In pre-Internet days, keywords were the basis of indexing new articles in published databases such as Current Contents. Keywords were critical to finding the relevant literature. Nowadays, keywords are more facultative. Internet search engines, freely available databases such as PubMed or subscription-based databases such as the Web of Science will locate relevant literature based on content anywhere in a manuscript. As more manuscripts are findable with given search terms and because different authors sometimes use different terminology for the same or similar concepts, researchers need to be selective in what hits they choose to read.

As an illustration, imagine that I am doing a keyword search on:

budworm *and* defoliation

Despite both of these terms being fairly specific, my Web of Science search turned up 458 articles published between 1969 and 2018. Plenty to go through, and much of it probably of little or no use for my purposes. Worse, I may have missed one or more key papers, since I am also interested in herbivory. When I did the following search:

budworm *and* defoliation *and* herbivory

I came up with only 43 references. And, when I did:

budworm *and* herbivory

there were 89 references. Thus, there were 89 – 43 = 46 references that had herbivory but *not* defoliation and 458 – 43 = 415 that had defoliation but *not* herbivory. All in all, there were 458 + 89 – 43 = 494 papers with defoliation *and/or* herbivory.[7] If I am interested in either of these phenomena, then either I will need to make the time to read 494 titles and possibly a large number of abstracts, or I will need to add one or more discriminating keywords to my search so as to pare down the number of references.

Ascending and descending searches

Imagine the following. You are reading an article that is central to your own work. You come across a number of cited papers in that article that pertain to your study, but which you have not yet read. You read these and find that some are useful, some not. Irrespective of whether or not you will cite these articles, you see that they contain citations to *additional* interesting references. Such *ascending tree searches* are particularly effective for exploring the past literature; the number of initial articles used for such a search magnifies—often considerably—the swath of what you finally find.

Having ascended, you will have tens and more likely hundreds of potentially relevant references. You can considerably increase the number of articles found by going the other way; that is, conducting a *descending* search. To do this, take a chosen reference and enter it onto a database tool such as the Web of Science[8] and see the papers that have cited it. Some source papers will have been cited hundreds or even thousands of times! This presents a logistical problem (or nightmare) since you can't possibly read all of the citing literature. Rather, what you can do is to add additional defining keywords to narrow the spectrum of interesting papers.

As you proceed through a descending search, your literature tree grows, but eventually the never-before-seen interesting papers dwindle. Such searches and associated reading can take huge amounts of time, even in reading paper basics (title, abstract). You will therefore need to be clever in narrowing the searches using appropriate keywords.

4
Avoiding Plagiarism

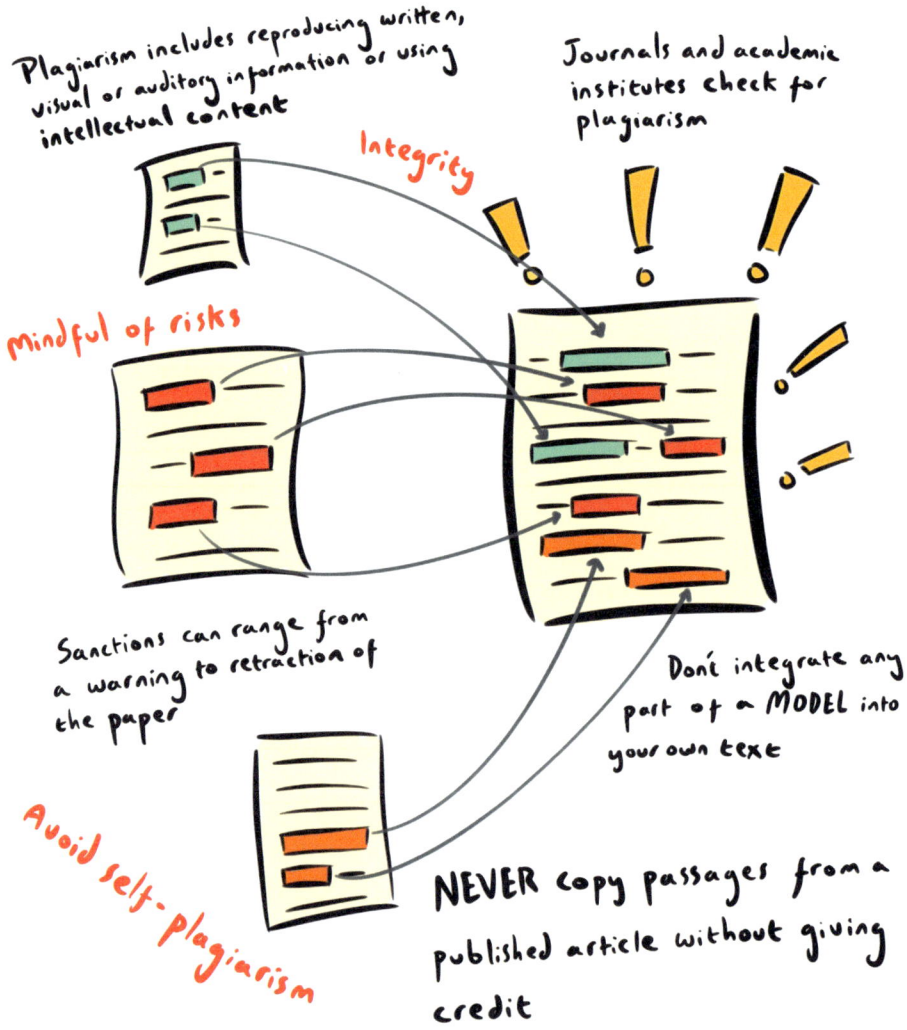

Scientists need to learn early that the use of intellectual content must be acknowledged. Plagiarism may occur inadvertently or intentionally. Regardless of the reason, plagiarism is a serious infraction. This chapter gives a brief account of different ways in which plagiarism occurs and how to prevent it.

Plagiarism is use of published material without due recognition of the source. Plagiarism can take two main forms: the reproduction of *physical content* and the use of *intellectual content*.

The first of these—copying word strings—is usually readily identifiable and checked by many journals and academic departments using increasingly sophisticated software. Duplicating a string of more than several words that appears in another published paper—irrespective of its intellectual significance—could trigger a red flag. If considered suspicious,[1] then it may be investigated further. The second type of plagiarism—not giving credit for the use of intellectual content—extends to ideas, findings or using published data. Intellectual plagiarism can be difficult to objectively identify because the suspected reproduction may physically differ from the source. These two types of plagiarism are evidently often not distinct: the verbatim copying of text (the original work) may also involve the wrongful appropriation of important intellectual content (the original idea).

Sanctions

Recognizing and acknowledging previously published content is usually taught before a scientist writes her first paper—but not always—and, either way, transgressions still occur.

Plagiarism may be penalized in several different ways. First, there is a social stigma, as it is tantamount to academic dishonesty. The details of the infraction can be spread (and be misrepresented) via social media. Second, a journal[2] or academic department may determine that plagiarism has occurred and act accordingly. Depending on circumstances this can vary from a simple warning to disciplinary action. Finally, authors or the holders of a copyright[3] may decide to pursue the alleged plagiarizer in a court of law. In academic publishing, the legal pursuit of infringements is complex, expensive and therefore uncommon; rather, if plagiarism is determined, then the article will be retracted by the publisher.

Be Mindful of Risks

Copying is not necessarily strategic—it can occur unintentionally, or intentionally, but out of ignorance that plagiarism is a problem. Unintentional plagiarism in particular is an ever-present risk for the simple reason that it's easy to copy and paste text. This may occur in two main ways.

The first is self-plagiarism—copying one's own previously published work. Imagine the following. I have published numerous articles on the same experimental system. Although the scientific questions differ from study to study, the system and core methods are identical. Should I simply refer to my previous published articles rather than use journal space to relate the methods for the Nth time? Many journals do in fact allow references to previous papers to replace methodological details, but what happens if an editor suggests or requires additional information? Should I simply copy and paste my previously published methods? If I were to do this, and assuming that I was not the sole holder of the copyright, I would risk copyright infringement. And would the other copyright holder(s) really react to this type of copying? Although there is potentially an issue for other copyright holders, there is arguably a larger problem: the social stigma. Scientists might judge my copying—even by me from my own past work—as unethical.

The second risk is in using *Models*. *Models* are published papers that writers can use to inspire their own writing. They will be covered in Chapter 7.

Imagine being a chief editor and receiving an email one day from the authors of a paper, published in a different journal.[4] The authors have noticed that a paragraph in a paper recently published in your journal was virtually identical to a paragraph in their paper. You examine the two paragraphs and your blood runs cold. Excepting just a few words, the two paragraphs are identical. How could this have happened? You contact one of the co-authors and he is able to trace the suspect paragraph back to its origin. He discovers—and now remembers—that he had copied the paragraph from the original *Model* paper, but forgot to distinguish the paragraph from his own writing. The *Model* paragraph was inadvertently integrated into the Discussion of his own paper.

Even if accidental, this is a candidate for plagiarism, both because extensive text was copied in its exact (or very similar) form, and since its intellectual content may have contributed to the new Discussion.

Beyond puzzlement, the original authors' concern could be that *others* who see the similarity between the two papers will wonder what happened. Therefore, much like the self-plagiarism scenario, the main issue is not that intellectual content is willfully taken, but rather "How and why did this happen?" and "What will others think when they see this?". The resolution of such an incident would likely be a published explanation and apology.

> COPE—the Committee on Publication Ethics[5]—promotes integrity in both research and its publication. COPE is a multi-faceted resource to guide and help editors with issues such as intellectual property, authorship, data reproducibility and journal management. Plagiarism and other ethical transgressions such as fabricating data can result in retraction of the paper in question and investigation of the researchers involved. Most reputable journals are now a member of COPE.

PART II
WRITING A GREAT PAPER

You have completed your study and are ready to plant your flag. You understand the significance of quality and scholarship to the scientific craft and have conducted an extensive literature review. All that's left is to write your manuscript and submit it to a journal.

Well almost.

There is no getting around the fact that writing and publishing papers is hard work. Creating and assembling a scientific study entails long hours, bumpy roads and occasional setbacks. But with the right preparation and method this is perfectly navigable and can even be enjoyable.

To start, you will have to gear up and plan your route—Chapters 5 and 6 will introduce you to the importance of your environment and how to organize and strategize writing. Even with the well-developed outline emphasized in Chapter 6, many beginning writers find themselves spending inordinate amounts of time, with little to show. Chapter 7 provides what is probably the most important tool to the struggling writer: Models. Models are quality-written publications that inspire your own writing.

With these basics, Chapter 8 introduces the structure of original research articles and provides guidelines to writing paragraphs and key sections. This will help prepare you to write your own paper, but as explained in Chapter 9, once done and published, it may not garner the attention it merits. The final chapters show you how to lure, hook and engage the reader (Chapters 9 and 10), and orient her to the most exciting findings (Chapter 11).

5
The Writing Mind-Set

Your working environment influences your writing. This chapter describes why writing is environment-dependent and how you can change or choose environments to make writing more productive and enjoyable.

Cell phone off—Internet off—*Almost* ready to write!

The mechanics of writing is all about opportunity and constraint. Take me. I am occupying a temporary workspace at home and have set aside several hours to make progress on this chapter. My new office will soon be outfitted, and I am waiting for the painter to arrive. Moreover, it's 8am and I have already planned videoconferences at 10 and 11am. Barring problems with the painter, this gives me just under 2 hours to write . . . and . . . the points to be discussed at the video meetings are in the back of my mind . . . I am even wondering whether—should the first meeting finish early— I will have some time to write before the second one. There are periods of concentration, moments of anxiety and fixed meeting times. Productive writing will be challenging.

Distractions are all too common. They can stem from personal imperatives or professional responsibility. They can often be managed. "Am I really expecting a crucial email this morning?" "Could I have organized my meetings so as to leave a larger block of time for writing?" Other distractions are in our immediate environment and seem to be the most trivial. Uncomfortable seating, noise and bad lighting are just a few of the countless derangements. Dealing with some of these is straightforward (adjust chair), but others can be annoying (noisy neighbors), sometimes making it difficult to write. But arguably the most severe distractions come from oneself. Writer's block, procrastination, incessant breaks. The symptoms can be anything from anguish to seeking refuge in something pleasant or easy to do, such as surfing the Internet or checking email.

Solving these opens the doors to effective writing, but *only if you really want to write*. If you are not somehow motivated, then no matter how you arrange your environment, you won't write. The following address motivational and environmental writing issues.

What Motivates You or *Could* Motivate You?

Motivation basically comes in two flavors. For some, manuscript writing is exciting and fun. We need no extra motivation, and even work happily despite external distractions. For others, motivation is more reactive than proactive. We don't enjoy writing and do it because we are responsible and meet deadlines. Improving the environment makes little difference.

Reactive writers sometimes think that proactive motivation is inaccessible to them. One either has it or doesn't. Nothing could be further from the truth. True, some people are easily motivated. But most of us who *are* proactive—and this includes myself—are not like this by nature—we are so because we work at it—we anticipate a goal.

Take a minute to consider these two sources of motivation:

- *Gaining expertise.* We learn every time we write. We learn about science and how to express it in written form. Consider the importance of learning more about your area of research, data analysis and interpretation, and the resolution of a scientific question. Should the writing go more smoothly than anticipated, can you learn why it worked and become a more productive writer?
- *Kudos.* Consider the feeling of achievement. Of making it to the summit. Do you enjoy the kudos coming from collaborators once you complete the writing and submit the paper? Celebrations when the paper is accepted for publication?

These motivators are not free or easy—they only work if achieving the goal *involves* work. The climb must be challenging to get some form of excitement and achievement out of it. If the path were a pointless stroll, then you wouldn't be doing it!

The keys to motivation are recognizing the milestones in your learning and in the manuscript as you climb, and realizing that reaching the summit is more than just being able to see for miles around. You will have built something very special, something new. You will get congratulations and can congratulate yourself.

Zero in on and Deal with the Distraction

Similar to contemplating and working on your motivation, assessing and improving your environment can positively affect your writing. And should your motivational issues *themselves* be environmentally related, then improving the latter can potentially transform you from a reluctant, non-productive writer into one who is enthusiastic and prolific.

The first thing you need to do is identify the source(s) of any distraction, and see whether or not it/they can be addressed. This can range from ameliorating any issues with the immediate environment, such as bad lighting or an uncomfortable chair, to dealing with the surroundings, for example, a noisy neighbor, or more distant, systemic problems such as street noise.

Not all distractions are straightforward to solve. Usually there is less you can do as the problem goes from the immediate environment (e.g. your chair, computer screen), to your surroundings (neighbors) and finally to the distant environment (street noise). Even dealing with one's surroundings can be unpleasant, such as asking an office mate to keep her voice down. Your motivation level is important here. If you are highly motivated and relaxed, then you may regard your disruptive neighbor as a minor encumbrance and cast a friendly look with a "shush" sign. If, on the other hand, you are unmotivated (and in a state of despair), then your neighbor may be the straw that breaks the camel's back. Don't let it. Take a few breaths and use diplomacy to solve the problem. Let this achievement boost your motivation.

Dealing with disturbance can be complicated. Some of us are uncomfortable confronting others, or fear that in doing so we might create an enemy. Sometimes the source is not easily actionable, such as corridor noise. How can you deal with such issues without going to the root of the problem? Consider three ways: compensate for the distraction, change your work schedule or move to another workplace. Compensating includes using headphones, moving or turning one's desk, or adding a plant or barrier to reduce visual disturbance. Either as an alternative or complement, consider altering your schedule so that you arrive well before and/or leave after the period of distraction. This may have the added benefit that office mates ask you "Why the change in schedule?", to which you can reply: "I work better before everyone gets in." They will get the hint. Finally, although possibly time-consuming and annoying, finding another workplace is sometimes the only way to get serious writing done.

Create Your Environment

Not all distraction comes from surroundings. Indeed, many stem from a need to create one's own space. Dusting the desk, shifting papers and books, making a pot of coffee, one last look at new emails, etc.—such rituals can indeed be just what is needed to write effectively.

But sometimes small adjustments are just not enough. There's a Friday morning discussion group that makes noise next door; I have to drop my kid off at the childminder Wednesday morning and, because of the long trip to my workplace, it would be better just to work at home; I have a videoconference writing session every Thursday and don't want to disturb my office mates. Such mundane constraints mean that we need to be flexible in when and where we write.

Moreover, a given environment may not be good for all writing needs. Different needs include casting the broad outline of a manuscript, writing rough or polished drafts, integrating citations, revising a manuscript, responding to reviewers' comments, etc. This is not to say that you should have a different workplace for each aspect of manuscript writing! Rather, some parts require more concentration, others more Internet work, and yet others space where you can discuss openly with colleagues.

A good writer has a good strategy set. These are discussed in the following chapters and include organization and actual writing mechanics and specific ways to structure a manuscript. But first and foremost a good writer knows where she writes best. So as to at least have a refuge from disturbance, she will establish two or more workplaces:

- *The office*. This is where you *should* be doing most of your writing. A PI needs to ensure that either your desk space is adequate, or special dedicated offices are available for writing.
- *The office at home*. This is for those fortunate enough to have a dedicated space at home, and to have periods when they can be away from their research institute. Although the home environment may let you escape from disruptions at your usual workplace, it can have the downsides of other types of distraction, such as your favorite books, television, roommates, frequent trips to the refrigerator, etc. Barring distractions, a home office is always useful before and after regular office hours and on weekends.
- *The café*. What could be better for reading and writing than having a continuous supply of your favorite beverage? Having a comfortable spot that is either quiet, or bathed in music or white noise, can be very conducive to writing.
- *The library*. Many prefer a more dedicated, academic setting, with the choice of either lots of open space or small, isolated booths. Libraries are places where you are sure to find silence and respectful behavior.
- *The meeting room* is a great choice when you write as a group, offering space for whiteboards, overhead projectors, and room for drinks and snacks.

Global Planning

The above recommendations concern reacting to and finding solutions to one-off or occasional distractions. Many disturbances however are recurrent, affecting many people and need to be addressed at the group or institutional level. An example is noise emanating from discussions in a neighboring conference room. Dealing with this on a case-by-case basis is senseless—a better approach being that those concerned raise the problem at a group meeting and come up with a durable solution.

6
The Start

Many scientists struggle with writing. The mechanics involved, the time spent, multiple revisions and waiting on collaborators to do their part. Scientists sometimes feel that their manuscripts do not do justice to the underlying effort and to the importance of the results. These troubles often stem from poor organization and a lack of strategy. This chapter presents the essentials of the writing mind-set and some of the basics in putting pen to paper.

Scientists are not taught to be good writers, let alone good science writers. They learn on the job, usually starting with the dissertation and interactions with their supervisor. Skills are honed and new ones acquired with each successive manuscript. Just like doing science, writing science continues to improve throughout a career.

Good science needs good writing. Conducting a monumental study has little value if you cannot communicate it clearly, accurately and convincingly. Writing is an essential link between you and the scientific community, influencing your impact, reputation and therefore your career. Moreover, good writing is beneficial because it requires that you think clearly about the science that went into your manuscript. There is no better way to gain a deep understanding of your own research than to write it up for publication!

We metaphorically climb a mountain every time we write a scientific paper. We start with a blank slate,[1] fill it in with what starts as primitive text and gradually attain a polished, integrated whole. Getting there means not only being able to write fluidly, but also integrate scientific knowledge, accurately recount how the study was conducted, display the results and then discuss their significance. But there is more to writing science than just recounting facts. Perhaps the most underappreciated facet is storytelling. Like fiction, a scientific article has a storyline, with a plot, characters, a beginning, development and an end. The science provides the bare bones of the story, but it's up to the authors to flesh it out and add the contours. You are the storyteller! Thinking about analogies between fiction and science in this way puts into perspective the challenges that scientists-as-authors face.

Yes, writing is work, but it can be enjoyable. We need to embrace the challenge and not view it as a mandatory exercise, but rather an opportunity to turn our *own science* into a written work for the world to see. A "work of science."

This chapter is about getting you to the starting line. To be a good writer you need to be prepared before even putting pen to paper—you must be organized and have a strategy.

The Importance of Time

Scientists are pressed for time. Teaching, meetings, grant applications, conferences, supervision and writing manuscripts. Some find it difficult to put aside just a few hours a week for writing. Writing time is often squeezed into gaps left by more pressing responsibilities. Others are unable to organize writing times in advance, and resort to binge writing or writing on the fly.

Fractionated, semi-planned writing works against quality. It results in errors, ambiguity and omissions. To see just why this happens, consider the following. If you had multiple chances at writing a manuscript, and the number of hours resulting in acceptable quality was 150 (which is not unreasonable), then what would have been its quality had you spent only 100, 10 or 1 hours? As we descend below 100 hours, small errors crop up—ambiguities, grammar, spelling, etc. Going down further, say below 10 hours, introduces both errors in reporting the Methods and Results, and speculation in the Discussion. The manuscript is shorter and more superficial than either the 150- or 100-hour versions. Below 1 hour, your manuscript would be reduced to a respectable outline.

Now what about 500 hours? Intuitively, the more time the better, correct? Perhaps more than 150 hours will be beneficial, but beyond a point, not only is more not necessarily better, it can be worse. Continued rephrasing and presentation tweaks make no difference to quality, and waste precious time that could have been spent on other endeavors. Practically speaking, writing for 500 hours will probably not happen in a block of three solid weeks; rather, it will be fractionated into months or even years! Collaborators will become disenchanted with delays and micro-revisions. They will have difficulties recalling the details of the study. Their contributions become more difficult to obtain and, due to waning enthusiasm, may not be of quality. To make things worse, you run the risk of being scooped by another research team.

Finding the "Sweet Spot"

Let's assume that you spend about 150 hours on completing a manuscript. How will you fractionate it? The extremes of a 1-week marathon or 15 minutes a day for 2 years are clearly non-starters. Between these extremes, how much time will you spend at each sitting and over how many weeks or months? Should you prioritize regular writing or rather accommodate writing as your already busy schedule permits?

You will want to write regularly enough to maintain continuity and so that your daily start-up routine is not too arduous. Writing regularly, at closely spaced intervals, also (obviously) means that you finish the whole project sooner. Spending 2 hours a day, 3–5 days a week is probably reasonable for most students and young faculty.

When you sit down to write, expect at least several minutes to gear up mentally. As you get better, start-up time will become vanishingly small. Depending on the task at hand and your resolve, you will want to write for *at least* a total of 30 minutes on any day, but probably be most productive at 1–2 hours per day.

You are not wedded to a particular number of minutes per sitting, sittings per day and weekly schedule. More demanding first-draft writing will require different organization than almost-pleasurable final revisions. Experiment with the main organizational parameters to find your sweet spot. More about fractionating writing below.

Pen to Paper ... Almost

With the logistics above and insights from the previous chapters in mind, you are about to actually write. Well, not quite. You need to do two important things first. You need a story and you need to organize how the story unfolds.

> *You Are Writing for a Journal.* Good science is the science, and it should never be influenced by the journal where your study is published. However, *the way* the study is written will be influenced by the journal you choose. This is particularly true for the Introduction and Discussion sections. Journals like *Nature* and *Science* have compact—sometimes single-paragraph—Introductions and Discussions, whereas papers in more disciplinary journals may dedicate several or more paragraphs to each. It is important that before embarking you have a good idea of the journal type targeted. See Part III for how to do this.

Your Story

Good scientific writing is first and foremost good story telling. This may sound surprising, because we usually equate science writing with strictly adhering to the mantra of rigor: accuracy, neutrality, clarity and precision. Good scientific writing *is of course* all these things, but there is little point in writing if readers are not interested.

Good story tellers are informative and captivating. They create intrigue. They know just how to satisfy a reader's curiosity. They reveal key insights but leave it to the reader to draw her own conclusions. Readers remember a good story.[2]

All good scientific stories share common features. Central is a fascinating mystery (the "puzzle"—which will have its own chapter). Others may have tried to solve the mystery (literature review), but have either failed or only partially succeeded (what your study addresses). You have a different approach (methods), and in applying it have found key clues (results). Your findings have clarified previously poorly understood facets of the mystery (discussion), but more work will be necessary to solve major parts and lead to a general explanation (conclusion).

There are many variations on how you can tell your story. Articulating it is both exciting and challenging. Exciting because here is your chance to explain *your* scientific discoveries to the community. Challenging since you do have latitude in telling your story, but not artistic license. Your story needs to lure and rivet readers without sacrificing the mantra of rigor.

Outline

A scientific article is a highly structured document. The writing needs to be accurate, of a high standard and polished. The idea that one can write a structured original research paper from top to bottom, section by section, paragraph by paragraph and sentence by sentence—without any kind of framework—is a myth. Actually, one *could* do this, but the paper would either fall short of its potential or, much more likely, be a mess.

Before embarking on such an important journey, you must first prepare the terrain. You need to outline.

The sunk cost fallacy. One does not simply sit, face a computer and start writing. Try and see how far you go. You may find that all appears to be going well. "This manuscript is a cinch!" But at some point—perhaps a quarter or halfway to completion—you see that the manuscript is not shaping up as you had hoped. The paper meanders. It's a dry, boring read. Given the time you've invested so far—possibly tens of hours—you find it *inconceivable* to start all over. You stick with your sub-par version, improving and accommodating problems that would no longer be an issue if you were to just start all over. This does not mean that the thinking and work going into an early abandoned draft will not contribute to the final manuscript, but rather that planning before actually writing will save considerable time and energy and produce a better paper in the end.

Outlining before you start to write serves several purposes. It makes you consider the major elements that will go into your paper. It allows you to get both a bird's eye view of the manuscript and think about the different strata of the contents. It is a written record that helps you avoid improvisation. Finally, and importantly, an outline is mutable: you can add, subtract and change as you develop the framework, and then as you write.

There are many fancy and sophisticated ways to develop an outline. Some writers benefit from these, but I have had the best results using a straightforward, classic approach. This views the manuscript as a series of logical suites—metaphorically like Russian dolls—with the largest being sections, within which are paragraphs and finally sentences at the finest scales.

> *Why?* When you begin outlining your manuscript and list potential paragraphs, ask yourself "Why this paragraph?" "What is it actually doing?" "Why is it here and not in another place?" I suggest that you write brief answers as notes in the margin and refer to them during the initial phases of outlining. If necessary, modify or move the paragraph as its function becomes clear.

The coarsest levels are easy. If you are writing an original research article, then this will typically be: Introduction, Methods, Results and Discussion (see Chapter 8). You then populate each of these sections with paragraphs or sub-section topics.

For instance, the Methods could be developed in a logical order as:

Methods

General culturing techniques;

Population measurements;

Experimental design;

Statistical analyses.

Each of these subsections would have one or more paragraphs, each of which developed a logical suite of elements. For example, Experimental design would become:

Experimental design

Experiment 1

Experiment 2

Details of each experiment are presented in one or more sentences. Before actually writing these, keywords could be listed and ordered so as to develop the details in each paragraph. For instance:

Experiment 1: initial population sizes, transfer methods (aliquot volume, vortexing, etc.), sample conservation.

This outlining procedure develops different hierarchical levels in a series of passes. The highest levels (sections) are immutable. You then go section by section and list the main components. Try keeping to these main points—there may be anywhere from a few to perhaps 10. You will need to list these in a logical order, and if you do not have experience doing this, see Chapter 7. Once you have a good draft of the main sections to the paragraph level, you go into each and list keywords or key phrases for sentences that will go into each paragraph. It is not necessary that you be exhaustive. Do your best at the first pass—you can come back in subsequent passes and add, change and delete these. It is a good idea to annotate your internal paragraph lists with observations of possible concerns or developments.

Write!

Now that you have attended to your environment, prepared your story and cast your outline, you are ready to write. Here are some of the basics for getting started, building momentum and monitoring progress (Figure 6.1).

Methods first, Abstract last. There is a logical order of how you should write your manuscript. Your Methods would have been detailed some time before you sit down to write the full manuscript. Start by writing the Methods, and do this as much as possible as the study is being planned and conducted. The next logical sections to write are the Introduction and the Results. Typically, these will be written in parallel, but there will be periods where you write the Introduction only, and others (as Results are finalized) where you concentrate on writing Results. Writing Results is usually much more straightforward than the Introduction, so some prefer to wait until the former is complete to start the latter. The final main section is the Discussion, which for many is the most difficult since it assesses findings, relates them to past work and gives prescriptions for the future. Once the writing is complete, you will have all the necessary elements to write the Abstract—the miniaturization of the manuscript.

Writing these sections is covered in detail in Chapters 8 and 9.

Your mission is to write. You have made a contract with yourself. Validating the contract means monitoring your progress. The time periods and word goals will depend on your experience, the technical difficulty in the writing and other

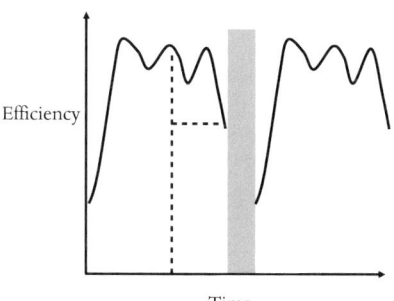

Figure 6.1 Hypothetical writing efficiency over two periods. It takes a few minutes to get near maximum efficiency. The decision to take a break (gray bar) can be based on a minimum writing period (vertical line) and continuing as long as above a minimum efficiency (horizontal line).

imperatives in your calendar. I usually target writing 250–500 words in two half-day periods of 1–2 hours each (my "sweet spot"). In initial drafts, I maintain acceptable quality standards, but these increase with each successive revision (see below). I always keep a basic log of observations, either at writing breaks or at end of the writing day. When the day is done, I assess my progress. Importantly, my primary goal is not words written, but rather *making progress*. I rank this on a scale from 0 to 3:

0 = No progress, little useful writing.

1 = Progress, but fewer words and/or lower-quality writing than targeted.

2 = Quantity and quality targets met.

3 = Quantity and quality targets exceeded.

Don't expect to be at levels 2 or 3 each and every day…unless of course you are already a very good writer. Being at level 1 *is progress*, but if your writing quality is not up to standard, then this will need to be improved in subsequent revisions. Do not get flustered if you are at level 0—what is important is to figure out why. Consider whether the issues come from your motivation, the environment or rather, from your ability to articulate ideas. If the latter, then the next chapter will be particularly useful.

It's important to take breaks. Writing can be anywhere from exhilarating to exhausting. Even if all is going as planned and you are making incredible progress, consider taking a break. As a rule of thumb, it's good to take a short (1–5-minute) break every 20 or 30 minutes.

Work with momentum. Some writing methods prescribe that the best times to take a break are at or near your peak in momentum. I would suggest continuing as long as you feel you have momentum. Work through your first break, and if you are still plowing forward as you come to your next break, then definitely take it, but reduce it to a minute or two. Consider extending each of your following work periods by a few minutes. Similarly, if your momentum is low (but you are still making progress), consider taking breaks a few minutes earlier.

You have the opportunity to revise and perfect. No one achieves the final version of a high-quality manuscript in a single draft. Incessant revising as you write will slow your momentum. Your main task is to write—you can revise afterwards.

Revise

Revision improves communicative quality and style. It includes deleting, correcting and introducing material. The need for revision varies depending on paper type (original research, reviews, commentary), author's writing experience and the journal targeted. Multiple co-authored papers tend to require fewer revisions than manuscripts with one or a few co-authors.

Many younger scientists are surprised to learn the number of revisions required to attain a high-standard original research paper. Typically, 10 or more revisions are necessary and it is not unheard of that a solo author will revise 50 or more times. Evidently, most of the work is in the first few revisions and, beyond about 10, the changes are mostly minor.

Some tips in writing and revising:

- Try to achieve a reasonable first draft. This does not mean laboring over the text, but rather putting in that little extra that makes it easier to revisit the paper.
- When writing the first draft, only revise as necessary. Incessant revision will break your momentum.
- Make notes about ideas, concerns and possible references as you write. You can attend to these subsequently in the dedicated revisions.
- When conducting dedicated revisions, attend to each section as a unit. Do not necessarily proceed section by section from the start to the end of the manuscript every time you revise.
- If you are co-authoring a paper, then alternate revisions with your collaborators.
- Let the manuscript "sit" for days or even a week or more between revisions. This brings a fresh view.

List it. A very simple technique that helps revising a paragraph is to hit ⏎ after each sentence. This makes sentences easily distinguishable as a list, enabling you to better see a paragraph's logical progression.

Your First Real Experience

Your first experience writing a manuscript will probably be as part of a group, and likely as follows. After months of data collection, analysis and numerous discussions, you are ready to begin manuscript writing. Your data are on spreadsheets, results in tables and figures, your bibliography includes hundreds of relevant articles, you have a computer complete with word processor, and graphics packages at your disposal. You have pinpointed several candidate journals for submission. The study is ready to be written.

There is no cut and dried recipe for how a collaborative group transforms a study into a manuscript. An experienced PI will be sensitive to the experience needs of her research team and the contributions each member can make to writing. Your first writing experience will probably be within your comfort zone. This may involve capitalizing on the contributions already made in conducting a literature review, or detailing the experimental methods or the results. It is possible, however, that you find that you cannot ensure your writing assignment. In such circumstances you should discuss the challenges as they arise with your PI. Encountering and resolving hurdles is a necessary part of becoming a better writer.

Communication and flexibility are key to group writing. This is not only true for deciding on and making your own contribution, but also in learning through feedback from more senior team members. Finding your "place" can be awkward. Each research group has its own culture, and—particularly when participating in writing for the first time—what is expected may be unclear. Don't get discouraged or anxious should more senior colleagues make executive decisions. Try to get clarification about your part. Your role as a group member will grow with time and you will likely increasingly contribute your ever-more valuable expertise to writing.

Collaborative Writing

Collaborative writing[3] is fun and enriching, but, being a multi-person endeavor, is not immune to surprises and delays. Scientific impasse, waiting for new results and writer's block all contribute. But because challenges differ between team members, and no two people react in the same way, differences in time management will invariably occur. This brings a social dimension to collaborative manuscript writing: one tardy (and unpopular!) author can hold up a manuscript for everyone.

Collaborative writing therefore needs a leader. Just like a symphony orchestra without a conductor, group writing without coordination will meander and waste time and energy. The leader—usually the PI—needs to make sure that each member of the writing team is achieving their assignments in reasonable time and at the desired quality standards. A good PI will account for differences in writing experience, and indeed young researchers can learn a lot by observing how a PI organizes and ensures effective team-writing.

Sometimes writing delays are unrelated to the manuscript at hand, and rather are due to changing priorities. A productive research team will have multiple projects in different stages of development. Efforts can rapidly shift from one project to another. A PI's decision to put a manuscript on hold can be trying for young scientists, who are looking to publish in timely fashion.

Despite the best of intentions, collaborative manuscripts are rarely on schedule. Be patient. Knowing your collaborators, their priorities and your own, and discussing scientific content, writing strategy and issues arising are most conducive to success, and making the experience both enriching and enjoyable.

A Writer's Toolbox

PIs and senior collaborators are important in helping to develop our writing skills, but ultimately, practical experience makes a good writer. Here is the set of guiding principles from this and previous chapters that will help you write:

1. *Motivate.* Think about your motivations. Why is this study exciting and important? Why is it important to *you*? How will it help your future research?
2. *Organize.* Develop the storyline, what is needed (figures, tables), who will contribute and the target dates you set to complete the first draft of each part of the paper.
3. *Outline.* Write a structured outline, remembering to add notes to different entries that may come in handy during actual writing sessions.
4. *Accomplish.* As you sit down to write, make bullet points of what you are going to accomplish during your writing day. Evaluate your endurance and progress by checking off these accomplishments.
5. *Don't revise* as you write the first draft. Searching for perfection will slow your momentum. Should important ideas emerge as you write, make a quick note of them and revisit later.
6. *Communicate.* Prioritize good communication between collaborators. Weekly meetings? Face to face, videoconference, telephone? Beware of discussing important points by email, as it can lead to regrettable misunderstandings.

7
Use *Models*!

A Model is a guide to structuring any part of a manuscript

choose from the top journals and the finest scientific writers

Avoid 'blind leading the blind'

Model → Manuscript

Select what you want YOUR paper to look like

Don't copy or plagiarize

Training wheels

Understand why a sentence or paragraph works

A Model is an article of the highest standard that serves as a stencil for writing one's own manuscript. Concretely, a phrase, sentence, paragraph, transition between paragraphs, overall layout of a section, table or figure all qualify as Models. A Model has earned its place: it has been worked, seen by external reviewers and editors, revised and made the journal's quality cut. Don't reinvent the wheel: search for success and emulate it. This chapter discusses the use of Models and walks the reader through real examples.

The Golden Rule of using *Models* is, in my view, *the single most important* guide to writing scientific papers. It's so important that it needs its own chapter.

Imagine the following. You go to a classical concert and are enthralled by *Syrinx* by Debussy—so much so that you purchase a flute and set out to play. Of course, you have no playing experience and find it a considerable challenge to even get a resonating sound out of the cursed thing. You try and try—different air displacements, distances from and angles to the embouchure hole...and finally, through trial and error, you pinch your lips in just the right way and lo and behold a sound resonates! You are elated and quickly assemble the flute and action the keys one by one while blowing. As a series of youthful notes emerge, you tell yourself: "*Now* I'm going to learn to play *Syrinx*!"

You have some talent and are able to figure out a short suite of notes from memory. But you are light years from the performance the other evening. You tune into an online video of the flute solo and watch it over and over again, sometimes in slow motion. Helpful, but you're still clunky. You decide you could use some basic technique and watch various flute lesson videos. You see, understand and review the bases that will enable you to play and learn further. It's a huge amount of work, but you steadily improve.

The flute videos served as *Models* that helped you achieve your own performance. You chose the *Models* based on your listening experience, the reputations of the flutists (that you checked online) and the quality of the performances/lessons themselves.

Models in scientific writing serve a similar purpose.

What is a *Model?*

A *Model* is a published article that inspires and aids us in our own writing. A *Model* has gone through countless revisions, reviewer comments, editorial and journal filtering. It has achieved a standard.

Concretely, a phrase, sentence, paragraph, transition between paragraphs, overall layout of a section, table or figure all qualify as potential *Models*. Metaphorically, a *Model* is like training wheels when we learn to ride a bike. We use them until we feel confident. But, unlike training wheels, we can return to *Models* at any time in the future—and without embarrassment!

Models are magic. When used correctly they guide the writer and enable her to see just how a section, a paragraph or a sentence can actually work

But just because an article is published does not make it a *Model*. Scientists are not all the same when it comes to writing quality, and journals can differ considerably in the importance they attach to filtering and correcting the scientific prose of the papers they publish.

It is therefore up to you—the writer—to identify those articles that both correspond to your needs and meet your writing standards.

Choosing a *Model*

To prepare for writing using a *Model*, you will need to do two things.

First, you will need to know what journal style-type you intend to approach. You don't want to find yourself emulating one style only then to submit to a journal with a different type. This said, most of your candidate journals will probably have minor contrasts between them. Sections like the Materials and Methods and the Results are generally standard and easily adaptable between candidate journal choices. Introduction and Discussion sections are less so.

Second, you will need to choose one or more *Model* papers. The *Models* will serve as "stencils," and can be used toward numerous objectives, including: thematic flow, sentence structure, emphasis and even the basis for choices in citations. You are not tied to a single *Model*, and more than one paper may serve to help writing of a given sentence, paragraph or section. Since casting an article can be very discipline-specific, I recommend choosing several papers on a similar or the same theme as yours.

My suggestion is that you ask yourself: "Which papers did I find clear, precise, informative and easy to read?" "What paper do I want *mine* to resemble?" (Note that we are talking about being inspired by scientific writing style and *not* copying or being influenced by intellectual content in one's writing.) Consider choosing both from papers recently published in the journal you are targeting, and from other publications that employ a similar writing approach.

> *The blind leading the blind.* Just because a paper is published in a prestigious journal and by great scientists does not make it a *Model*, and you can be unwittingly doing yourself a disservice by selecting a poor model to emulate. Look for writing that is clear, active, flows logically and is not wordy. You will know it when you see it, but it is still helpful—at least in the beginning—to get the opinions of more senior colleagues.

Just a few pointers before starting:

- Let *Models* help, but not constrain. Don't turn to *Models* for each and every sentence!
- Beware of using more than one *Model* for a given paragraph or section of your paper. Different *Models* will have at least subtly different writing styles and by using more than one, you may be creating what is akin to Frankenstein's monster!
- *Do not copy content!* The stencil concept teaches you to infer the logic and style of presentation, *not* to plagiarize.

Models ≠ Plagiarism

As discussed in Chapter 4, plagiarism is the replication of specific information originating from a published source without giving due credit. The *Oxford English Dictionary* defines plagiarism as:

The action or practice of taking someone else's work, idea, etc., and passing it off as one's own; literary theft.

All plagiarism involves the copying or using material that has some degree of originality and can be traced to another source. The use of a *Model* is different: it is a writing strategy.

> *The five-word rule.* There is no foolproof way to determine all instances of plagiarism. A group of observers may agree that a string of three words in one instance is plagiarism, and eight words in another is not. A good guideline is to adhere to the five-word rule; that is, never copy a string of five or more words from any published document. Nevertheless, if a particular short expression is important to your message, then copy it, place it in quotes and provide reference to the source.

Let's Start!

Now we are ready. You decide to embark on the Introduction. You have read Chapters 5 and 6 and have set up your environment and outlined your paper. You open the mostly blank word processing document and…bang! You don't even know how to begin.

Now pull out your desired *Model* and read the Introduction, paying close attention to the logic of flow and the style. Ignore the science. It may take you anywhere from 10 minutes to perhaps an hour to feel confident that you see and understand what makes the *Model's* Introduction work.

Number each paragraph in the margin. On a separate document—the document that will become *your* Introduction—do the following:

For each numbered paragraph in the *Model*, state on your document what it generically accomplishes. These are your (temporary) paragraph headings.

For example, "1. This paragraph presents the puzzle of explaining cooperation, especially in humans." "2. Presents different mechanisms that may underlie cooperative behaviors in humans." "3. Presents empirical evidence in support of some of these mechanisms." "4. Relates a contention to the relevance of what some claim to be the central mechanism." "5: States the purpose of the study, how it is executed, and its main findings."

Next, for each of these headings, list statements of what is done or accomplished. You may want to do this sentence by sentence. Take, for example, paragraph 2:

2. Presents different mechanisms that may underlie cooperative behaviors in humans.

Sentence 1 (S1) states that a number of mechanisms have been identified

S2 says that that they can be divided into three basic categories (types 1–3)

S3–S4 present type 1 and how it works

S5, presents type 2…

S6, presents type 3…".

Do this for each paragraph of the Introduction, and then examine the overall structure.

> **Example**
>
> The first paragraph of the *Model*[1] reads:
>
>> Cooperation is a pervasive phenomenon in biological systems, and despite considerable study, its establishment and maintenance are incompletely understood. Theoretical work starting with Hamilton's seminal papers identified a number of key features that promote cooperation. While many empirical studies have tested theory using social insects and cooperative birds and mammals, a growing number have employed microbes, given their rapid evolution and experimental control relative to metazoa. Predators and parasites may either be the basis of social behaviors, such as cooperative defense, or constitute a cost that potentially impacts other cooperative behaviors (e.g. resource access and sharing, quorum sensing). Such costs may differ between individuals adopting different social behaviors, and include energy or time committed to defense or resistance, or costs associated with trade-offs involved in evolved resistance to enemies. Despite their ubiquitousness in nature and demonstrated importance in population ecology and evolutionary biology, the impacts of natural enemies on the ecology and evolution of microbial cooperation remain largely unexplored.
>
> Here is a skeleton of this paragraph:
>
>> The paragraph gives a brief historical overview of cooperation, then focuses on work on cooperation in microbes, and then on the effects of predators and parasites on microbial cooperation.
>>
>> *Model* sentence 1 (M1) describes the overall importance of the topic, and yet it is not well understood.
>>
>> M2–3 say that the theory is well developed, and empirical studies abound, many of them more recently on microbes (transition).
>>
>> M4–5 say that natural enemies such as predators and parasites may be direct or indirect causes of social behaviors and gives generic examples.
>>
>> M6 says that natural enemies are generally important, but impacts on microbial cooperation are largely unknown (this is the teaser).
>
> Now let's use the *Model* to write the first paragraph of your paper. Your study is on herbivory and host plant population dynamics. There are few apparent similarities between the *Model* and your paper but remember that the *Model* is a guideline for logical and stylistic support, and nothing more. At each step of the writing process verify that you are telling *your* scientific story, and not reproducing the intellectual content of the *Model*.
>
> So here we go!
>
> M1 Cooperation is a pervasive phenomenon in biological systems, and despite considerable study, its establishment and maintenance are incompletely understood.

My first sentence (S1): "Herbivory is a ubiquitous phenomenon in ecological communities and, despite considerable study, its effects on plant populations are not well understood."

Notes: I could have written the end phrase in several different factual ways. In choosing "effects on plant population dynamics are not well understood" I am revealing a fact about these systems; that is, I am not creating this as a mirror to the Model. The longest string shared between Model and manuscript is four words and is perfectly generic.

> M2 Theoretical work starting with Hamilton's seminal papers identified a number of key features that promote cooperation.

S2: "Empirical work has shown that plants can limit herbivore damage through the production of defensive chemicals."

To remain factual, I had to deviate from some of the structural analogies in the Model. The "how herbivores" is important and will be elucidated in one or more subsequent paragraphs. Like the Model, my sentence is very general. I will verify its wording later in one of many revisions.

> M3 While many empirical studies have tested theory using social insects and cooperative birds and mammals, an increasing number have employed microbes, given their rapid evolution and experimental control relative to metazoa.

S3: "Herbivores may be affected by plant chemicals in different ways, including altered feeding and dispersal behavior, arrested development and increased sensitivity to parasites."

Like before, to remain factual and in continuity with my previous sentence, I have deviated from a 1:1 analogy with the Model. I don't want to introduce different taxa here, but rather focus on effects. I may decide at a later point in the Introduction to include one or more specific systems.

> M4 Predators and parasites may either be the basis of social behaviors, such as cooperative defense, or constitute a cost that potentially impacts other cooperative behaviors (e.g. resource access and sharing, quorum sensing).

S4: "Defensive chemicals are however costly to produce, resulting in slower growth and lower germination rates than the non-defensive chemical-producing wild type."

The Model refers to contrasting effects of predators and parasites. Instead, I want to go directly to costs, since these are a central feature of my study.

> M5 Such costs may differ between individuals adopting different social behaviors, and include energy or time committed to defense or resistance, or costs associated with trade-offs involved in evolved resistance to enemies.

S5: (Skip)

I decided to skip any analogy here and go directly to the final sentence (the "hook"; see Chapter 9) of the paragraph.

M6 Despite their ubiquity in nature and demonstrated importance in population ecology and evolutionary biology, the impacts of natural enemies on the ecology and evolution of microbial cooperation remain largely unexplored.

S6: "Despite the pervasiveness of plant defensive chemicals in nature, the role of their within-population variation and its underlying causes in plant–herbivore population dynamics remain unexplored."

This is an important sentence. Indeed, the last sentence of the first paragraph often sets the tempo for the paper.

The complete first draft paragraph reads as:

Herbivory is a ubiquitous phenomenon in ecological communities, and despite considerable study, its effects on plant populations are not well understood. Empirical study has shown that plants can limit herbivore damage through the production of defensive chemicals. Herbivores may be affected by plant chemicals in different ways, including altered feeding and dispersal behavior, arrested development and increased sensitivity to parasites. Defensive chemicals are, however, costly to produce, resulting in slower growth and lower germination rates than the non-defensive chemical-producing wild type. Despite the pervasiveness of plant defensive chemicals in nature, the role of their within-population variation and its underlying causes in plant–herbivore population dynamics remain unexplored.

This would not be the definitive paragraph—I will read it over carefully and revise it for accuracy, possibly add, delete, emphasize or de-emphasize different points, and add references.

The Exercise

Now that you have seen the basis for how *Models* work, let's make them work for you. Consider again the flute analogy at the beginning of this chapter. *Models*, such as face-to-face instruction or videos, will help you improve your playing technique. This is the same for writing papers, but with the important difference that *Models* can help you practice your writing technique or even inspire the final paper itself!

Here is the most important exercise of this book. All you need is:

- To have at least embarked on a research project and, preferably, be ready to write a manuscript.
- A *Model*.

The key is the choice of the *Model*. It needs to be written to a level to which you aspire. I suggest asking your PI or a senior colleague for suggestions.

1. Pick one paragraph from the article. If this is your first time doing the exercise, then choose the first paragraph of either the Introduction or the Methods. Copy and paste the paragraph into a text editor. Use the return key to align the beginning of the sentences of the paragraph on the left margin, like this:

   Model Sentence 1

   Model Sentence 2

   Etc.

2. Read the sentences one by one, paying close attention to their structure, logical flow and what they achieve. Ignore the scientific content.

3. Now, read the sentences from the last to the first, paying attention to how each sentence is predicated (or possibly not) by those above it.

4. Once you are at the top of the sentence list, go back down the list sentence by sentence and add brief notes under each regarding what made the sentence "work."

5. Now go back up to sentence 1, and at the end of it, hit the return key so that there is space for your first sentence.

   Model Sentence 1

   My Sentence 1

6. Transpose the structure (*not* strings of words themselves, nor the content!) of the *Model* sentence into your own research. See the previous section for examples of how to do this.

7. Continue down, transposing sentence by sentence.

This whole exercise should take no more than 15 minutes. Be fluid. Once the paragraph is complete, it is time to revise so that it is coherent and accurate, both scientifically and with respect to the context of your study. The logic should flow seamlessly. These revisions should take about the same time as writing the first draft.

Try doing this exercise again with another *Model*, but this time do not adhere so closely to the *Model* structure. There may be phrases or sentences that you do not emulate, and others that you include in your paragraph (to ensure the logical flow) which were not inspired by the *Model*.

8
IMRaD

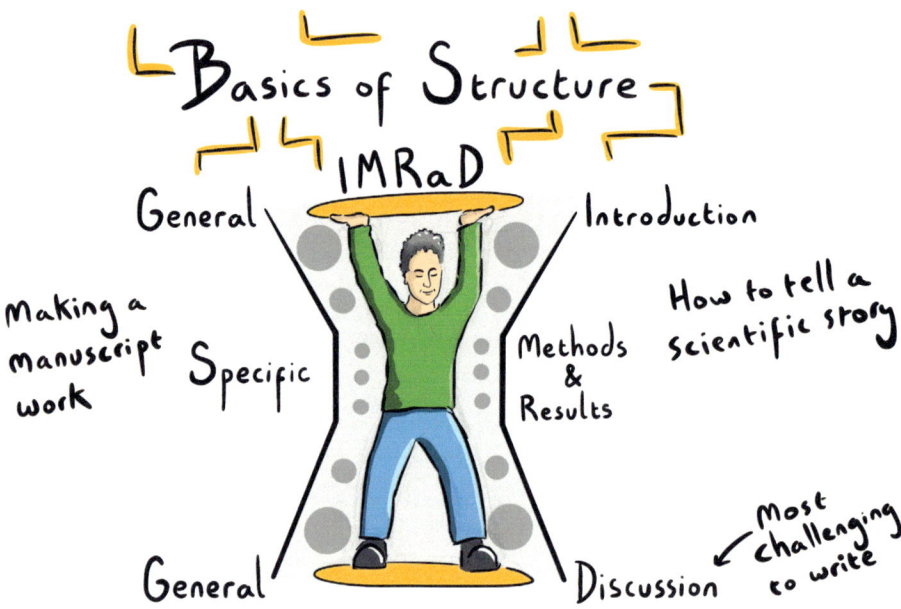

Original research articles almost invariably follow a time-tested structure—IMRaD—Introduction, Methods, Results and Discussion. This chapter presents the structural basis of IMRaD and emphasizes the most challenging section to write—the "D."

IMRaD—Introduction, Methods, Results and Discussion—is the most commonly encountered structure for telling a scientific story. The familiarity of all scientists with IMRaD facilitates communication. IMRaD *always* progresses in the logical series: past, question, present, answer, future. This chapter won't rehash what has been so well said about IMRaD in many books and on countless Internet sites. To drive home the essence of IMRaD, I want you to do the following exercise before reading any further.

Exercise 1

List the types of information that go into a manuscript. For example, the experimental design, statistical analysis, the conclusions. Try to be specific enough so that your list comes to 10–15 entries, and that one or more fits into each of I, M, R and D.

Now take a few minutes to think about how your own research maps onto each entry in your list and how the entries are interrelated. For example, if you indicated "experimental design" in the Methods section, then first think of the basics of your experiment, and then consider how this relates to entries in the Introduction, produces specific Results and possibly influences points in the Discussion.

The point of this exercise is to see the logical sequence of entries and their interrelations. You likely listed some or all of: I (previous study, unresolved questions, particular question of interest, hypothesis), M (methods to test hypothesis 1, 2, 3…), R (findings 1, 2, 3…), D (interpretation, relation to previous study, conclusions and future directions). It should be plain to see that IMRaD is a template for writing a study according to the scientific process.

The hourglass

The metaphor for the IMRaD structure is an hourglass (Figure 8.1). The Introduction begins wide and general, narrowing to the specific puzzle. This narrowing doesn't change much in the Methods and Results, and finally widens in the Discussion. IMRaD basically says: this is what is out there (wide), this is what we don't know (narrowing), this is what we did (specific) and this is what it means (widening).

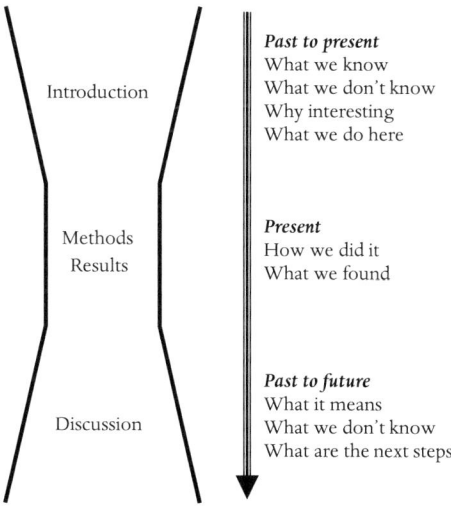

Figure 8.1 Hourglass structure. The main flow is top to bottom, with later parts addressing problems posed in earlier parts.

Although there is considerable flexibility in the proportions of the hourglass, two guidelines are useful. First, the hourglass should never get too wide (generalities), the maximum usually being a few lines at the beginning of the Introduction and in the Discussion. A manuscript starts with the scientific background, leading in logical steps to a question in the final paragraphs of the Introduction. Then there is a sudden break to how the current study actually addresses the question (Methods) and what was found (Results). The Discussion explains the findings, their broader relevance and where to go in the future.

Second, the proportions in the Methods and Results sections will depend on how general and extensive the study is. Methods and Results usually have an internal substructure that gives the hourglass an undulating pattern in the middle. For example, you may have conducted laboratory experiments, statistical analyses, and computational modelling, and each of these will start with generalities (e.g., general culturing techniques), followed by more specific methods and details.

> **Exercise 2**
>
> Select two articles from your *Models*. Begin reading the first, not paying too much attention to details. As you skim IMR and D, picture the width of the hourglass changing. On a separate piece of paper trace the hourglass paragraph by paragraph. Now do the same for the second article (and eventually other articles, should you have time), tracing the hourglass next to the first one. Did the hourglass shape reflect the scientific and communicative quality of each study? Your interest in each?

Paragraph structure

The Introduction and Discussion are the two most challenging sections to write. Nevertheless, their paragraphs have a characteristic internal structure. Each starts with a statement of what the paragraph is about. This is followed by one or a series of developments in a logical order that can be chronological, categorical (e.g. laboratory followed by field studies) or may lead to the subject of the following paragraph. The last sentence usually "ties the knot" (e.g. summarizes, concludes, generalizes, resolves…) and/or creates a transition that will be taken up in the first sentence of the next paragraph. Similar to the overall structure of a paper, each paragraph in the Introduction and Discussion typically has an hourglass shape.

The Introduction

An Introduction will have all of the following components[1]:

1. The general context.
2. What is known: review of the literature.

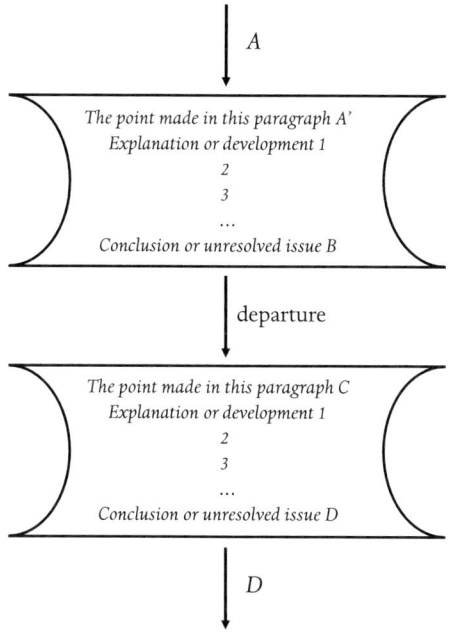

Figure 8.2 Generic paragraph structures in an Introduction and Discussion (the overall logical flow is represented by subject points A–D). The first sentence (and internal development) follows from the content of the preceding paragraph (indicated by '), or marks a departure from the preceding paragraphs.

3. What is not known: the puzzle.
4. Why the reader should care: why the puzzle and its resolution are important.
5. What the study does: test a hypothesis or address a question.
6. The actual approach taken.

Do the following exercise that will help you learn the essential features of an effective Introduction.

Exercise 3

Using the same *Models* from Exercise 2, carefully examine the structure of the Introduction. See how and where the authors provide general context, what is and is not known (and whether they provide detail and/or refer the reader to referenced material), lead up to a question, problem or puzzle and their hypothesis, how they articulate the interest and importance, and finally the description of the study's objectives and how these were accomplished. Make a note of the logical flow of the paragraphs and development within each paragraph as per Figure 8.2. Are some of your *Models* more effective overall or in particular parts of the Introduction? Why?

The Introduction lures the reader into a scientific question. Stimulating the reader's curiosity is so important that it needs its own (the next two) chapters. The Discussion is no less crucial and is the focus below.

The Discussion

The Discussion is the most open-ended section in an original research article. It is an update of the Introduction, with the added challenge of interpreting the Results and presenting future directions. Discussions take on a variety of forms, both in terms of basic layout, and the fine structure.

The overall design of the Discussion can differ from journal to journal. It can be a separate section called "Discussion" or "Conclusions," or may consist of one or more paragraphs tacked on at the end of the Results section. Some journals adhere strictly to the section name and what the section should achieve, whereas others are flexible, the only real checks coming from the views of referees and editors.

Why is the Discussion so important? After all, the Introduction has provided the background, the basic puzzle, the question being addressed and a trove of relevant literature. The Results presented the findings—the astute reader can draw her own conclusions and their relevance by referring back to the Introduction. Indeed, one *could* stop there and have learned the essentials. What the Discussion adds is relevance—something that the reader would be thinking about anyway while reading the Results, but which the authors will actually provide, and likely do so in much more depth.

Discussion Last, But Not Quite

The Discussion is the final written section. You will probably have been contemplating it from the beginning and possibly will have taken notes as the writing of the earlier parts took form. Despite considerable latitude in writing the Discussion, it needs a plan, a foundation. Fortunately, you need not look far. The IMR is the foundation to build the D.

Think about it as follows. A scientific study has three main periods. The past up to the present (I), the present (M and R), and the past to present to future (D). Whereas the first two periods and their sections follow on from one another, the third—the Discussion—traces back to points in the Introduction, goes through the present (M and R) and projects into the future. Thus, a scientific paper is not a simple, unidirectional logical line from beginning to end. It's a loop.

A good Discussion revisits the Introduction. It puts both the general topic and specific question addressed into perspective. It provides answers and generates new questions.

The List of Six

An effective Discussion will present the:

1. problem;
2. resolution;
3. contributions;

4. limitations;

5. conclusions;

6. implications.

> **Exercise 4**
>
> Select one of the articles from the previous exercise. Read the Discussion and note where you find each of the six above points. Think about how these structural features contribute to the effectiveness of the Discussion as a vehicle for closing the study. If the *Model* has fewer than all six points, consider how the absence of each affected your assessment.

Writing a Discussion

Before presenting examples of the key first paragraph in a Discussion, here are some basic pointers for when you actually start writing.

First, list the six points for an effective Discussion at the top of your document and develop each in a sentence or two in the context of your study. Try to include keywords, phrases and descriptions that will enable you to return to the list and transform it into scientific prose. It is possible that you may want to downplay some of the six points.

Second, transform the points and descriptions above into an outline of the paragraphs in your Discussion. You can start with a title and a one- or two-sentence description for each paragraph. Place the paragraphs in a logical order, eventually using one of your *Models* as a guide. Note that you may have one or more paragraphs relating to each of the six points, and that some of the six will be treated in the same paragraph. You can always come back and add, delete or modify elements from this list.

Third, although there is no "one size fits all" for how to organize content, consider the following parameters: scale, importance and confidence. Scale refers to going from small to big, local to global and few to many (or the reverse). Importance answers the question "Why care?". Confidence is the extent to which your results are robust and general.

Fourth, if in outlining your Discussion multiple themes emerge and you decide to develop each in some detail, consider establishing subsections with explicit titles. Not all journals allow this. These subsections should form a logical sequence. *Models* will be useful since this is a more advanced feature of Discussion writing.

Fifth, embark on the actual writing. Proceed in a series of "sweeps," starting with notes and observations of what you plan to include, why, possible concerns, etc. With each sweep, you will need to populate your text with supporting references. Some of these citations will be discussed in some detail, others will serve as support for claims or statements. As you proceed, highlight parts that need more work and resolve past concerns.

Keep in mind that writing the Discussion takes time and effort, but is highly rewarding. It's where you inform the reader of why your findings are interesting and important.

The Crucial First Paragraph

The Discussion is special—it is the only section where the first paragraph *really* matters. The first paragraph of the Introduction usually presents general context—not very important. The first paragraph of the Methods is a strictly factual recount (or overview) of what was done and how. The start of the Results is somewhat constrained to follow the logic of the scientific process, and is not arranged a posteriori from most to least significant findings. The first paragraph of the Discussion however is pivotal: it announces the interpretation of the science just presented and presages the paragraphs that follow.

So, let's look at some real examples (with cited references omitted).

> Reiterate results
>
> We have presented and demonstrated the use of several of the more common methods for dealing with uncertainty in model parameters when designing reserve networks. These efforts have provided four key results. First, all of the approaches explicitly allow one to quantify the effects of uncertainty on model results and the consequences for subsequent decisions made in the face of uncertainty. The ability to quantify the impact of uncertainty creates flexibility in the management decision process and provides a mechanism for describing realistic outcomes. In some cases, stakeholders may be willing to take greater risks if such risk creates the potential for a larger pay-off; in other cases, stakeholders will demand that any action provide the greatest amount of certainty in results. However, under the precautionary principle these decisions should be based on worst-case scenarios. Increases in persistence probability pay the price of requiring greater certainty in key model parameters, and this tension is quantified in these uncertainty models.
> Second, distance…
>
> (*Ecology Letters*, vol. 9, pp. 8, 9)
>
> *Comment:* This is a straightforward and powerful way to launch a Discussion. It is most effective when the study is complex, with multiple important results. This is not to say that it is a good idea to dedicate a whole paragraph to each and every significant result. Rather, the question you need to ask is whether the *first* paragraph should announce that you are about to describe your main results in more detail than in the Results section itself.
>
> Relevance
>
> The series of tests we performed supported niche mechanisms as the primary drivers structuring these prairie grasslands, even though the observed species

> abundance distribution at CDR was qualitatively consistent with neutral theory. These results illustrate that a distribution fitting approach could lead to incorrect assumptions of neutrality. Additionally, not only can niche and neutral models predict similar species abundance distributions, but the distinction between the relative fit of alternative predicted distributions to species abundance distributions is contentious and depends on which goodness-of-fit measurement is used. Our ZSM fit outperformed the log-normal fit in terms of χ^2, but not R^2, which also suggests the need for better niche models that incorporate species traits such as R^* with dispersal limitation and stochasticity, and methods to distinguish between the relative importance of these processes. Fitting a neutral distribution to observed species abundance data was proposed as a means of assessing the importance of neutral processes in ecological communities; deviations from this null expectation should indicate greater relative importance of other processes. Because our observed species abundance distribution was consistent with neutrality while our stronger tests were not points out the potential limitation of using neutral distributions as a null expectation and test of neutral processes. We reassert that stronger tests are needed, including tests for relationships between abundance and ecologically relevant traits.
>
> (*Ecology Letters*, vol. 9, pp. 19, 20)

This paragraph is amazing. It is a microcosm of the study itself and, although it reads in some respects like an Abstract, it is far more informative. The sentences and their wording are very carefully chosen to create a logical sequence of information highlights, balancing communicative quality with scientific detail. The reader almost does not need to read on, but wants to, because this lead-in sparkles.

Context first

> Over the past century, grasslands and other seminatural plant communities in temperate Europe have suffered dramatic decline in their area due to land-use changes, and thereby once widespread vegetation types have become highly vulnerable. But what we have lost in area we may still have in species richness. Many grassland plant species with long life cycles, slow intrinsic dynamics and relatively large populations appear to occur as remnant populations and communities in modern landscapes. Given the biological traits of the species, it may take a substantial length of time before the adverse consequences of habitat loss and fragmentation become apparent in terms of greatly reduced local and regional species richnesses.
>
> Our results strongly support the notion of remnant grassland plant communities with a slow response to environmental change…
>
> (*Ecology Letters*, vol. 9, pp. 74, 75)

This is a highly effective first paragraph, taking the tack that the scientific problem is important and speaks for itself. But presenting context alone in a first paragraph has its risks, since the writer may get bogged down in the details of recontextualizing the Introduction. To make this Discussion take off, the authors create the effective "rupture": the contrast between what appears to be an almost insurmountable problem ("...it may take a substantial length of time before the adverse consequences of habitat loss and fragmentation become apparent...") and the reporting of an interesting finding at the beginning of the second paragraph. This device is similar to the "hook" used in the Abstract and Introduction, and described in Chapters 9 and 10.

Strong, short statement

> Our sample of species included trees belonging to different evolutionary phyla (conifers vs. broadleaves), with different water transport characteristics (ring porous, diffuse porous and tracheid bearing), leaf ecology (evergreen vs. deciduous) and intensity of management practices (native unmanaged stands vs. intensively managed plantations of introduced species). Because of this wide variability, the consistency of our results suggests a degree of generality and may reflect the overall effects of 'extrinsic' factors such as plant size on water and carbon relations.
>
> (*Ecology Letters*, vol 8, p. 1186)

This is a clever and highly effective first paragraph. The authors employ three devices. First, the paragraph is short and to the point—only 79 words. Second, the first two-thirds of this paragraph relates the robustness and generality of the study. Third, although avoiding detail (which will come in later paragraphs), the authors clinch the paragraph with the main result in less than 10 words.

The above examples of effective first paragraphs are representative of the different strategies you may use in beginning to cast this important section. In general, avoid taking a passive tone, rehashing a multitude of details and over-emphasizing limitations. This goes for the first paragraph and indeed the Discussion as a whole.

Just to drive these messages home, below I give a reworded example of a first paragraph, based on an actual manuscript rejected (for a variety of reasons) from *Ecology Letters*. The Discussion began:

> Figure 8 showed $Z0/Z1<6$ for most species, suggesting that Minit was close to or above Mopt. X did not follow this rule with $Z0/Z1>6$..."

Would anyone want to read further?

Although there is no right or wrong choice in structuring the first paragraph of the Discussion, the launch you choose will influence the tone and logical progression for the remainder of this section. Because of the importance of the lead paragraph, it is a good idea to write first drafts of two or more versions, and to outline logical follow-on paragraphs for each alternative. Compare these and decide which is the most effective communicator.

The ending

Like any good story, a scientific article will have a purposeful ending. The ending may be one or a few sentences, a paragraph, a dedicated subsection or a full section (e.g. Concluding remarks).

Endings have many purposes, including:

- tying up loose ends;
- reiterating the main findings;
- presenting greater relevance;
- paths for future work;
- guarded speculations;
- limitations and their potential resolution.

The article's ending affects the reader. It metaphorically talks to the reader on the mountain peak: reviewing the climb, how it could have been better accomplished, what is its greater significance and future climbs in store.

9
The Vitrine

Scientists are reluctant to read articles without good reason. The first places many look are the title and abstract but also—importantly—figures. These three elements encapsulate a study. This chapter provides recommendations for writing these often-underrated parts as well as outlining what makes a good figure.

The craft of scientific writing is both technical and stylistic. The technical part requires an understanding of the basics of expressing science in an accurate and neutral way. Technique is the logical organization of a manuscript, from the overall layout to the paragraph-by-paragraph sequence and finally the composition of each paragraph. Style is the story and its narrative. Style must never compromise the mantra of rigor: accuracy, neutrality, clarity and precision.

Potential readers are largely indifferent to all this. They have to decide whether to read your article in the first place.

Let me explain.

Before the advent of email, the Internet and routine use of media in accessing scientific publications, interested readers would consult paper copies of journals directly, or thumb through periodic volumes of abstracts, such as Biological Abstracts. Whatever way, tens or even hundreds of new, potentially relevant scientific articles were available for reading every month.

Nowadays with the Internet, not only do we have articles at the tips of our fingers, but we have more of them. Total numbers published increase by several percent every year, which may seem paltry, but if you had to deal with 500 relevant new publications last year, it will likely be around 520–550 this year.[1] Add to this the previously published papers, some of which you are still discovering. These have accumulated over time and are still scientifically (or at least historically) valid. Depending on how specific your area is, the total relevant literature is probably in the range of 1000 to 10,000 published papers.

Simply keeping up has become—and is—an ever-growing problem. Depending on what one wants from a study, it can take anywhere from tens of minutes to several hours to read an article. But as we saw in Chapter 3, busy scientists can't spend 10 minutes, let alone an hour, on every paper of interest. Rather, they have no choice other than being choosy. Through experience they learn what to choose.

Clues about the content of a scientific article stare the reader in the face. These are the title, abstract and figures. In seconds you can get a bird's eye view and—if interested in more detail—read on to understand the main findings in just minutes. Thus, for readers to come to appreciate your technique and style, they first have to be lured by what they see in the vitrine.

Titles, abstracts and figures are special. They are encapsulations of the manuscript. The title is the headline: the place where the authors can advertise. A title can be anywhere from dry and purely factual, to clever and alluring. The title says "Hey, this is why you should stop and look!". In contrast, the abstract is a depiction of the whole manuscript. Authors have some flexibility in how the abstract is written, but not as much as for the title. The abstract says "This is what our study is all about." Figures visualize and advertise the main findings while rigorously keeping to the facts. A good figure both clarifies complex interrelationships and creates new insights. Figures say "This is why our study will dazzle you."

The reader may stop or continue anywhere along the way in the logical sequence: title, abstract, figures. She may spend seconds on a title and at most 5 or 10 minutes on the abstract and figures. And nothing says that a reader convinced by a title need read the abstract and explore the figures. A smitten reader may very well go straight to the Introduction.

This chapter is about elements that help scientists decide whether to read a paper. Knowing these elements will help you cast more alluring papers and attract readers. I begin with what *all* potential readers read: the title.

Titles

I present and comment on a series of titles based on fictitious study of a real insect pest—the spruce budworm (*Choristoneura fumiferana*).

> A field study of the population dynamics of the spruce budworm *Choristoneura fumiferana* and its natural enemies

This title gives several keywords/key phrases (field study, population dynamics, spruce budworm, natural enemies), and expresses some level of generality (a field study of population dynamics), a specific type of interaction (host–parasite), a type of tree (spruce) and type of damage to the tree (buds), and a particular species (*Choristoneura fumiferana*). Even without any knowledge of the paper, the title tells us that it is discipline-specific, possibly to forest entomology or population ecology.

The following could appear disparaging, but it's not. Our fictitious study might very well pass quality control with flying colors and be published, but not be read by anyone outside of this particular system and subject. This may be the authors' intention and it's a laudable one, but some authors either are not aware of how much flexibility they have with the title, or are, but do not think it makes any difference. It can.

Different readers have different preferences. Some prefer dry facts and are even put off by overly active or showy titles. Others, especially non-specialists, are looking for how your study may yield general insights of interest to them. Clearly, no single title can win over all potential readers, so in writing a title, authors need to have some idea of the segments of the readership they wish to attract.

There are three main dimensions in constructing a title.

1. *Specificity–generality*. The title at the beginning of this section is fairly specific relative to this alternative: "Herbivore–enemy interactions and their implications for host plant dynamics." If we believe that our results are sufficiently general to apply to other defoliating insects, or for that matter, to any herbivorous arthropod, then the new title is a definite candidate. This would attract a wider range of readers, but possibly lose some specialists. A more general title does not necessarily mean a net gain in readership.

2. *Emphasize Introduction/Methods or Results/Conclusions*. The classic, descriptive, baseline title at the beginning of this section presents what the study is, but not what was found or concluded. An example of how the title can be changed to emphasize a result is "Natural enemies impact forest tree dynamics through indirect species interactions." The decisive term "impact" gets attention. An interested reader might read more about what this impact actually is from the abstract, or perhaps skim through the manuscript looking for a key figure, or even decide to go directly to the Introduction.

3. *Plain–artifice*. The V1.0 title is plain and descriptive. It will attract a segment of scientists. Nevertheless, if we wish to bring in reluctant readers then adding some artifice is an effective way. Remaining factual is not inconsistent with using terminology, phrasing or word plays as devices to attract attention. Consider:

"Tri-trophic cascades: how natural enemies indirectly influence forest tree dynamics." The authors use the catchy and accepted term "tri-trophic cascade" to emphasize the mechanism behind their main finding.

Note the different way that each of these three facets attracts (or would possibly repel) segments of the potential readership. A title can be situated most anywhere in this three-dimensional space. What strategies are effective for coming up with the "right" title given the plethora of possibilities?

Let the title brew. Start by writing possible titles before or shortly after you begin writing the manuscript. List them on the first (title) page, eventually with the name of the co-author who suggested each one and any comments. As the manuscript progresses, add, modify, promote and demote (but don't delete) titles, until you finally settle on the winner when submitting the paper.

Get everyone involved. You may be surprised who has the best ideas for titles. Encourage all co-authors to submit ideas. Consider discussing the final choice as a group.

Use Models! Read recent titles in your shortlisted target journal(s). Like the use of *Models* for writing the main body of a manuscript, avoid copying content, or simply copying and pasting phrases.

Do not sacrifice. In choosing a title, take care not to choose phrasings that either inflate or lessen the scientific quality, interest or importance of your work.

Number of words or characters allowed. Look on the journal's website to see the limits for the number of characters and words allowed in a title. Sometimes this can constrain the set of possible alternatives.

Abstract

Many authors view abstracts as not being worthy of special attention because they are "simply a summary of the paper." "It's best to write it quickly—most readers won't use it anyway."

Wrong.

The abstract provides a synopsis for those who are either potentially interested in reading the complete story, but are not sure, or have already read the paper, but want a refresher.

A good abstract:

Miniaturizes the manuscript. The abstract should mirror the main sections of the paper: Introduction, Results and Discussion. Sometimes it is opportune to include basic Methods. As a rule of thumb, begin writing by dividing the abstract among the main manuscript sections in roughly equal proportions. You can then adjust the length of each to obtain what you view as the most effective package.

Reads easily yet has information to "chew on." The abstract should entice a range of prospective readers. Some journals have separate summaries for non-specialists.

A good abstract minimizes complex notions and jargon for readability, but stimulates curiosity.

Has a "hook." The "hook" is carefully chosen words, a phrase or a statement that elicits curiosity. The hook pulls the reader into the study through some kind of paradox, counter-intuition or mystery. The hook's home is the abstract, but may also appear in the Introduction. Consider the following example (with the key words deploying the hook emphasized in bold): "…Some authors have suggested that not just confidence but overconfidence—believing you are better than you are in reality—is advantageous because it serves to increase ambition, morale, resolve, persistence or the credibility of bluffing, generating a self-fulfilling prophecy in which exaggerated confidence actually increases the probability of success. **However,** overconfidence also leads to faulty assessments, unrealistic expectations and hazardous decisions, so it **remains** a puzzle how such a false belief…" (*Nature* 2011, 477, pp. 317–320).

Seamlessly links different parts of the manuscript. A good abstract has a train of logic with smooth bridges between the different sections.

Is the right length. Don't feel that you need to go up to the full abstract word limit. Should you exceed it, think carefully of how you can prune and reword.

Below, I present—sentence by sentence—a short fictitious abstract relating to the title at the beginning of this chapter.

Sentence 1: One of the most spectacular phenomena in forest ecosystems is the cycling of lepidopteran defoliators.

> I've exploded out of the gates by qualifying the context of my study as "spectacular." This was done with some care since I said "*One* of the most…". The rest of the sentence suggests that the study is probably oriented toward a fairly specialized audience: forest ecosystems; lepidopteran defoliators. I am being intentionally general in saying "cycling of lepidopteran defoliators." Does this apply to all such species? Some? Many? As written, the phrasing is generic, and makes no claim that all such species are associated with cycling. If I wanted to be more specific and assuming that at least several species do cycle, then I would qualify this phrase with "many." If I wanted to be more cautious, then I would use "some."

Sentence 2: Although natural enemies such as predators, parasitoids and pathogens impose significant mortality in these herbivores, their role in population cycles is largely unknown.

> Here's the "hook"—the mystery. These species kill their hosts or prey, but it is not known whether this is sufficient to generate population cycles. Note that I said *largely unknown* and not just *unknown*; this suggests that there is some evidence (possibly circumstantial) linking the enemies to population dynamics. The mystery therefore

is not a complete surprise, but it reflects the facts. N.B. I am now a bit uneasy about the use of "spectacular" in the first sentence. Once I finish writing this draft of the abstract, I will revisit "spectacular" and see if it is better to attenuate it so as to match the temperature of the hook. I will also verify that the use of the hook in the abstract coincides with its use in the Introduction (and not use the exact same wording between the two). Note too that if I were to go over the abstract word limit, then I could have written just "natural enemies" and waited for the Introduction to explicitly cite "predators, parasitoids and pathogens."

Sentence 3: We identified patterns in parasitism from a museum collection of 11,332 juvenile spruce budworms, *Choristoneura fumiferana*, collected over a 72-year period.

These are the Methods in a nutshell. Because only a single defoliator species was involved, I decided to give its Latin species name (but this could have been deferred to the main text). Also note that I included two numbers. Neither was necessary to state here, but because they are impressive—reflecting the robustness and generality of the method—I give them here. The actual manuscript will possibly contain tens or hundreds of numbers, but I decided that these two stood out. Note that neither number is actually a result—should there be a number emerging from a striking result, I would consider including it, but want to be careful not to oversell the study.

Note also that Sentence 3 reduces the spectrum of possible natural enemies, because (as would be explained in the main text) only parasitic ones could be identified. It is not completely clear from this sentence whether the 11,332 juveniles were *all* parasitized—I attempted to clarify this by using "from a museum collection" as opposed to "in a museum collection." Moreover, the main text will be specific about what is meant by the general term "juveniles"; there is no point in naming larval instars or imagoes here. Finally, the reader will wonder how on earth the authors were able to sample these insects over a 72-year period. The qualifier "museum collection" will stave off any misunderstanding.

Sentence 4: Parasitism rates varied considerably between natural enemy species and over time, but showed a significant correlation with temporal patterns in juvenile numbers.

I would have preferred to start this sentence with "We...", but refrained since the previous sentence began with this same word, and it's difficult to see how using the first "We..." could have been avoided. Moreover, starting point-blank with the result "Parasitism rates..." is a more active voice.

This sentence does two main things. It says that science is not perfect: the results were variable. The statement also gives the main result of the study, possibly key to enticing the reader into the main manuscript.

Sentences 5, 6: Temporal autocorrelation analysis showed that peaks in parasitism followed peaks in host density with a delay of 6 years. When fitted to a population dynamic model for the system, this delay produced the observed 30-year period in outbreaks of this forest pest species.

> This is a straightforward presentation of a second set of main results. It mentions two scientific approaches: statistics and mathematical modeling. The reader would need to delve into the paper itself to see the answer to the (hopefully burning) question of the sensitivity of the main finding to different values in model parameters.

Sentences 7, 8: Our results are consistent with natural enemies playing a role in spruce budworm outbreaks. This has implications for the use of parasites in biological control programs.

> We end with a short statement that addresses the main question posed in the study, albeit using the factual term "consistent with," rather than a stronger qualifier. A short concluding sentence closes the abstract, and the reader will need to go the Discussion to learn more about the possible applied importance of the study.

This abstract has 150 words.

Figures

Figures are the graphic representation of data and results, usually produced using specialized computer software. Whereas methods underlie scientific quality, figures contribute to communicative quality. Good science and effective communication are necessary for your paper to be read and remembered.

Figures should be reserved for the most important, interesting, informative and impressive features of your paper. Figures catch the reader's attention: wasting a figure on an obvious or unimportant result sends a message that the results which did not make it into figures are even more inconsequential.

The most effective figures have two central qualities:

Value added. A great figure produces insights that are difficult to communicate in words alone. The saying "a picture is worth 1000 words" applies to an excellent figure.

Intermediacy. A good figure is neither overly simple nor hopelessly baroque. The best figures have one or a small number of key messages.

Figures are often among the first elements that a prospective reader examines, particularly when short of time. A great figure can compel even hesitant readers to look closer, and so needs to be carefully chosen and designed. Most journals limit figure number, which does not preclude going over the limit and relegating some to electronic appendices. In such cases, you will need to decide on the figures that go into the main manuscript "vitrine." Often, we reserve these special spaces for conceptual background, experimental set-up and of course, the major results.

A good figure is self-contained and legible. The reader should not have to go to the main text to understand the basic features of the figure. The axes should be clearly labeled. If symbols are used, then it is best to have English terms next to them, or at least in the figure legend. Any points, lines or shading should be labeled either in the corner of the figure, or in the legend. Make sure that details like the size of typeface, and points, lines and shading are legible and easy to distinguish.[2]

Two-Minute Colleague Test

Show a fairly well-worked version of your figure to a colleague. Include the figure title and legend. Give your colleague 2 minutes to examine the figure. Then, ask the person what the figure shows, and whether she had to struggle to understand it. If your colleague had problems understanding the figure and its main messages, then consider how your figure and its legend can be improved.

10
The Puzzle

A study may be of the highest standard and provide new answers to a scientific question, but yet is still ignored. A great paper has an interesting puzzle that hooks the reader. This chapter discusses how to engage the reader in the quest to seek pieces of the puzzle.

Let's briefly take a step back before proceeding.

We have underscored the importance of converting science into manuscript. Deciding it is time to plant your flag is the *first* step. You are ready to sit down and write. Adapting your environment—both your mind-set and your surroundings—is the *second* step. You have established the conduit between the science and your word processor. You have it all at your fingertips: the literature review, the big data set, there are unexpected results, you have questioned previous theory and your study opens the road to novel research avenues. You have all of the specifics to support each milestone, have outlined your plan in the IMRaD framework and are prepared to use *Models* as necessary—this is step *three*.

Now you need to tell your story. You need to communicate and, in doing so, draw the reader in.

Communicating has several dimensions. The most fundamental is the mantra: accuracy, neutrality, clarity and precision. Readers need to understand what you are saying—otherwise there is no point in saying it. After this, a scientific paper needs *context*. What do we know? Why is it important? What is missing? The majority of your readers will know (far) less about the context of your study than you do. Next, the paper needs *emphasis*. Which of the results are the most significant or surprising?

Readers need guidance. Readers need a mystery.

Need a Mystery

The literature survey you will have done provides much of the foundation for your write-up: the general state of affairs, specific debates, recent progress and outstanding questions. Although this is essential to writing a good paper, you need more than just the bare facts to write a *great* paper. What is required to make a good paper great?

You need to captivate the reader with a mystery. Your story needs a *puzzle*.

A scientific puzzle is an unexplained observation. An interesting scientific puzzle is complex, but not complicated. Investigating whether high temperatures kill some organism is not an interesting puzzle. Asking why a population goes extinct over a particular temperature range even though individual survival is not affected is an interesting puzzle.

A good puzzle evokes counter-intuition, mystery, intrigue and suspense. A good puzzle challenges the reader and stimulates her curiosity.

The puzzle can be formulated around:

- A question—it's answerable.
- A theory, hypothesis, prediction—it's testable.
- A contradiction, controversy, paradox, problem—it's resolvable.

A scientific study can be based around one of many alternative questions, problems and predictions. So, choosing *a* puzzle is straightforward, but identifying *the* puzzle involves some work. If you are fortunate, a leading thinker in your area would have recently articulated an interesting problem that can form the basis your puzzle.[1] More usually however you will have to investigate and decide for yourself.

Consider the following scenario.

You have completed a series of experiments aimed at determining how a population of bacteria adapts to a range of resource substrates. One of the experiments tested the specific question of whether a key gene is involved in such adaptation. You sequenced and identified the genes responsible and followed their frequencies in the population during the course of the experiment. In analyzing these data, you noticed an unexpected result: novel bacterial mutants emerged that were able to actually metabolize the waste products excreted by the primary adapting lineage. This turns out to be the single most exciting result in your study. Thus, as originally planned, the theme of your study is population genetics and evolutionary biology, but now there is an ecosystem twist to it. How might this affect the articulation of your puzzle?

Let's look at two approaches.

Original intended question. Usually, the puzzle would be articulated around the question of why you conducted the study in the first place. So, for the above hypothetical example the question may be: "Are the genes involved in bacterial adaptations to novel substrates conserved?". Unexpected results don't change the puzzle, but rather the excitement of the study will shift from the Introduction to the Discussion, since the major result was not anticipated in the beginning. This said, some authors might preview the theme of the surprising result in the Introduction.[2] In the bacteria example, this could take the form of a couple of sentences dealing with the actual functions of beneficial mutations.

Targeted audience. What if your results appear to be of more interest to ecosystem ecologists than to population geneticists? Or theoreticians rather than experimentalists? Again, as above, the science must remain true and complete. Thus, you would have difficulty pitching the puzzle (and the Introduction and Discussion more generally) to ecosystem ecologists if only one (striking) result is relevant to this audience. This could, however, be an option should the key result have led to additional experiments, thereby rendering it an object of investigation, rather than just a novel result. In such a case in the bacterial example, the original mutant results would be reported as a prelude to extended studies following the (reported) observation of secondary metabolite use. Developing the puzzle without creating a straw man requires care, and writing the manuscript for a broader audience evidently means acquiring the literature background to do so, and may also affect your choice of journal.

Need to Care

In carefully articulating the puzzle, you would have surely thought, "How am I going to explain the importance of this to the reader?". Will what you as author think is important and exciting be viewed likewise by others? Given the diversity in readership is there really any point in trying to write with likely readers in mind?

Indeed, there is. Readers are influenced by the science, transmit this influence in their own publications and recognize the source of the influence through citations. An engaged reader is more likely to do all of the above compared with one who struggles seeing the point.

Despite high-quality science, some articles captivate readers, whereas others are a chore. What explains this? True, a good article has the essential ingredients of a well-developed context and an interesting puzzle. But a great article does more: it captivates the reader to the very end. An engaged reader gets involved in the story and cares about the outcome.

There are several complementary approaches to engaging the reader:

Developing the history and evolution of the puzzle. Based on your literature survey in Chapter 3, provide an account of how the puzzle emerged and has defied resolution.

Creating curiosity. We saw the use of the "hook" in the abstract in Chapter 9. This same device—a series of observations punctuated by a surprising statement—can stimulate curiosity in the Introduction.

Maintaining suspense. As results unfold, briefly refer to pieces of the puzzle that fall into place, and new questions that are generated.

Presenting relevance. Discuss how pieces of the puzzle affect how we view the whole.

Opening new doors. Present how this study creates new interesting questions for future research.

Exercise 10.1

Choose two or three of your *Models*. For each, identify the "hook" and parts of the article that use the five above approaches to engage the reader. Are the papers that you found most exciting when you first chose them to be your *Models* those that use most or all five devices to involve the reader?

We know a good paper when we see it: excellent science, clear presentation, interesting and important findings. A great paper needs more. It needs a story that creates mystery in the beginning and engages the reader to the end.

11
Emphasis and Finesse

Scientific writing should be rigorous but also needs to avoid monotony. The judicious use of adverbs can greatly enhance communicative quality and reader engagement. This chapter presents strategies for writing interesting prose without sacrificing neutrality.

Writing science is the same as storytelling, but with one important difference. Whereas the storyteller is free to invent, create and even deceive, the scientific writer is a reporter of facts. The neutral basis of scientific writing does not preclude some degree of freedom in style and expressing points of view, and indeed reading science would be *horrible* if pared down to the bare, dry facts.

As previously stated, the mantra of rigorous scientific writing is accuracy, clarity, neutrality and precision. As a reporter of facts, you are transcribing information, which presupposes both an in-depth understanding of your area of study, and lucidity and logic in how to express and organize it in your manuscript.

However, introducing the style that makes your story more interesting risks compromising accuracy. You may still write logically, but in being too liberal with qualifiers, you could mislead and confuse the reader. Style is therefore a balancing act of engaging the reader without sacrificing accuracy.

Qualifiers

This book is replete with adverbs. Many, most, some, very, extremely, etc. We use adverbs to describe quantities and qualities when we don't have an exact number or precise description. Adverbs come in all shapes and sizes depending on what is factual and whether we wish to be conservative or to emphasize. For example, if there were 52 observations of something out of a sample of 100, then this could be correctly qualified as "most" or "the majority," or more accurately as "a slight majority." We could also refer to 52 as "many," which removes the notion of more than half; 2 of 100 is also technically "many," but more acceptable would be "some" or even better, "few."

> *The Sorites paradox* or the "paradox of the heap" refers to the use of indeterminate quantities like "some" or "many." Fifty-two, 54 or 98 observations out of 100 are all "majority," and we can confidently say that 98 is a larger majority than 54 or 52, but would be at pains to use different terminology for 54 vs. 52.

Another example. If in conducting a statistical test I find that the probability p of a Type 1 error is 0.001%, and the norm for calling a result "significant" is any p less than 5%, then do I refer to my result as "Significant"? "Highly significant"? "Incredibly significant"? In my field of population biology, the first two qualifiers would be acceptable: the first is a conservative qualifier and the second less so. The third would be viewed by any reader as inappropriate terminology.

I intentionally overdid the above examples. Both had a single number (52/100 and 0.001%), but each had more than one possible qualifier. Which to use would depend on desired emphasis. There are bounds however, and these are shaped by norms in what appears in published articles. Consider two different types of norms. The first type is the vocabulary actually used to describe a phenomenon. For example, one would *never* qualify statistical significance as "highly substantial significance," although if this were to go unnoticed by peer reviewers and be published, readers would understand what you mean. The second type is the regularity with which a qualifier is used in the literature. For instance, just surpassing the halfway point (52%) may be most commonly referred to as a "slight majority." "A very slight majority" is also terminologically acceptable, but less used.

Issues with Qualifiers

Authors provide findings—readers seek them. A clever author can use emphasis to point the reader to what she believes are the most important and interesting results. This relies on the author's own judgment, given several issues that may compromise the accuracy of a manuscript.

Exaggeration. Excepting checks by peer reviewers, there is nothing really stopping authors from emphasizing too many results or exaggerating certain findings. Qualifiers need to be dosed both in their frequency of use and degree. For example, if you were to start reading the Results section to find the first four sentences beginning with "Interestingly,…"; "Surprisingly,…"; "Counter-intuitively,…; and "Interestingly,…" (again), you may question the authors' motives.

Dear Prudence. Writing is almost complete and it's time to "plant your flag." You are convinced that your study has made a true, important discovery. The first line of your Discussion reads: "This is the first study to…" I suggest taking a step back and being realistic: it is impossible to verify with certainty that your study is *the* first. A more modest and likely truthful claim would be "To the best of our knowledge, this is the first study to…" or "This is among the first studies to…"

Consistency. Emphasis should be used consistently. For example, if you refer to a statistical p value of 0.001 as "highly significant," then you should use the same association throughout the manuscript. Qualifiers of degree such as "some" and "many" do not refer to specific quantities, but may have relative differences ("some" is fewer than "many"). They should be employed as consistently as possible.

Limited information. Qualifiers may be misleading when information is limited, as for example calling 2 out of 3 "many" or "most." Best to call this "2 out of 3."

Weasel words. The weasel word qualifies a claim with no evidence given. A reader may accept the claim without verification, verify the claim or be unable to verify it. Weasel words include: many, some, potentially, generally, mostly… For example, a statement like "Studies generally show…" without references to the claimed studies, misleads the reader. Words that portend to reveal facts without support should be avoided.

Terminology and Jargon

Terminology is the use of specific words to facilitate communication. Different disciplines and subdisciplines have terminological norms that scientists need to learn if they want their work understood by others in their area. Jargon contrasts with terminology in being more technical and more school- or subdiscipline-specific. Whereas terminology is necessary for naming important objects, jargon is often shorthand for frequently used (but not necessarily important or universal) objects.

Terminology and especially jargon can create communication issues, including:

- Different people using the same terms for different or contrasting objects, or different terms for the same object.
- Overuse of terminology resulting in a large, unwieldy lexicon. A non-specialist who struggles with ambiguous and unfamiliar terms is likely to stop reading early on.

Journals may explicitly request that authors minimize the use of jargon for the obvious reason that readers (especially those outside of the area) will have difficulty understanding the contents. A young scientist needs to learn to be accurate in the use of terminology, use it sparingly and avoid jargon.

Differences between Original Research, Reviews, Commentaries, Opinions...

As we saw in many of the previous chapters, science is not written according to a single, rigid set of rules. There is considerable wiggle room so long as the scientific writing mantra is respected. Different disciplines, journals and even research groups differ in writing style. This is particularly apparent between article types. Not only are stylistic differences in original research articles allowed, but they are expected in Commentaries and Opinions.

Authors write to be read. They need to think both like a writer and a reader. In reading an original research article, I would value a clear, concise exposé. I want the writer to use qualifiers sparingly. I learn particularly from good figures, because I'm busy and don't want to spend too much time reading details (which can be relegated to supplementary information). For a Review article, I want the big picture and major unanswered questions. I need the authors' help in explaining intricate and complex concepts and findings, and putting them into perspective. Finally, a Commentary or Opinion should tell me what is so special that I will be interested in learning more. The author's personal perspective is the hallmark of these article types. Chapter 27 discusses non-original research articles in more detail.

Crying Wolf and *Relief*

Emphasis should to be used where just and effective.[1] No matter how wonderful the results, using qualifiers like "exciting," "unexpected" or "highly" sentence after sentence can actually harm an article. As mentioned above, the reader eventually fails to see differences in emphasis, becomes circumspect and even irritated.

Although findings will stand on their own without qualifiers, the dry results of statistical tests do not tell readers about scientific importance. Whenever we highlight a result we are signaling to the reader: "I realize that you will not remember each and every result, but *this one is really* important and interesting." A star-studded study will emphasize only the most meritorious results, whereas an unremarkable study will highlight all of the important few.

Metaphorically, the landscape of a scientific article is most effective when there are valleys, hills and a few mountains. From the French, it needs *relief*.

Selective emphasis can take several forms:

Single (poignant) qualifiers

The universe of emphasis and expletives in science is large, yet respects the mantra of rigor. Terms like "wow" are definitely out. A word like "amazing" implies something fairly unique, impressive and surprising. "Amazing" is not a scientific qualifier, but could be used in a more informal context, for example, in a Commentary.

Qualifiers vary from the subtle "note that" or "for instance," to the highlighted "surprisingly," "unexpectedly" or "counter-intuitively." Qualifiers can be modulated, such as, for example, "quite surprisingly," "highly unlikely," etc.

> *No proof.* Unless your study involves mathematics, avoid using the terms "proof," "proven" or "proves." Strong or conclusive evidence is not proof.

Emphasis through wording

Emphasis is not only adverbs—it can also be done through wording. The wording of a phrase, sentence or paragraph can be altered so as to emphasize (or deemphasize) a finding or concept through a logical argument. Again, similar to single qualifiers, the mantra must be respected.

Emphasis through wording is—not surprisingly—more challenging to apply than single qualifiers. It is usually employed in the Introduction and the Discussion sections, but depending on the journal, can also be in the Results. For example, the italicized part of the following sentence emphasizes a result: "*Consistent with the hypothesis that parasitic wasps differentially parasitize budworm larvae on spruce trees*, we found that budworm damage to fir trees was greater than to spruce trees (ANOVA $F = 13.43$, $p < 0.001$)." The relevance of this finding could have appeared in the Discussion only, but gives the result more impact by appearing alongside it.

Visual material

Tables and figures are standard tools for packaging complex or complicated information. But they are also vitrines: they direct the reader's attention to particularly important, clear or exciting results. As related in the previous chapter, figures in particular are a major means of emphasis.

Journals typically limit numbers of figures and tables, but nothing obliges you go to the limit! Like adverbs, unnecessary entries saturate the reader. Authors need to be choosy in what they select to go in the vitrine. Balance the visual effect with the importance, surprise, clarity and robustness of the result. So, for example, your key result may be that Y increases exponentially with X. If you have a large number of data points that are fitted to a curve that clearly demonstrates this finding, then a figure may be warranted—and all the more so should the result be unexpected. Thus, you are saying through your figure: "This is a clear, simple, important *and* robust result, look at the statistical fit to *all* that data!".

Journal cover

Only 20 years ago, journals were exclusively published in print, and many still do maintain a print version. Despite electronic publishing, journal covers are as alive as ever. Getting the cover of the journal is a major accomplishment, attesting to the interest of your paper, and is likely to increase its readership and impact. Although some journals exclusively solicit cover photos, others have an option to propose candidates, either when submitting the manuscript or when sending the final accepted version.

PART III
CHOOSING WHERE TO PUBLISH

You now have the tools to write a great paper and have started putting these to use. The next important step is choosing where to submit your work. How do you choose a journal?

Journals are black boxes for beginners and the experienced alike. Knowing how journals function educates about how they select science—your science. Chapter 12 demystifies journal function, what editors look for in a paper and how they get it. Chapters 13 and 14 examine in more detail how journal gatekeeping works and what this means for writing your manuscript (Chapter 13), and what to expect from a well-run journal when submitting your manuscript (Chapter 14).

With this knowledge you are ready to choose a target journal for your paper. But there is more to this choice than only the editorial board and the articles they publish. You need to understand the world of academic publication: non-profit and for-profit publishers, paywalled and Open Access journals, costs paid by subscribers or by authors, journal rankings, and journal articles, preprints, edited issues and books. The world of publication is undergoing a quiet revolution to make science as transparent and accessible as possible. The most familiar feature of "Open Science" is "Open Access"—publication being freely accessible and reusable to everyone. Chapter 15 discusses how Open Access is impacting choices, and how scientists are reacting to issues in publishing by posting their manuscripts on preprint servers. Deciding whether one wants to publish Open Access is only one of many criteria in choosing a journal, and Chapter 16 helps navigate this landscape.

12
How Journals Operate

Ask a scientist—particularly at the start of their career—how journals work and you will either get a blank stare or a logical but inaccurate reply. The only people who know the details of journal workflows are managing editors and chief editors. This chapter introduces the reader to the basics of journal function, what journals look for in a manuscript submission, and how they apply protocols of selectivity while maintaining fairness.

Journals are black boxes. A submitted manuscript goes in, and some time later a decision letter comes out. It is easy to imagine the basics of what goes on inside the black box. Someone does an initial screening; if the manuscript passes, then the paper goes to a specialized editor who does further screening, who, in turn may recommend desk rejection or be sufficiently impressed to send the paper out for external review. The reviewers eventually submit their reports, the editor makes the publication decision and sends it to the corresponding author.

Without any doubt, the most frequent question during my tenure as chief editor at *Ecology Letters* was "How do journals *really* work?". My reply would typically be "Just as you would *want* journals to work!". The curious colleague would then give their

An Editor's Guide to Writing and Publishing Science. Michael Hochberg, Oxford University Press (2019).
© Michael Hochberg 2019. DOI: 10.1093/oso/9780198804789.001.0001

account of how they thought journals were run. Although most people got the basic inner workings right, they invariably missed the most salient feature:

> Journals put everything at their disposal into identifying, improving and publishing choice science in the fairest way possible.

This chapter unpacks this credo.

Journals in a Nutshell

The modern-day journal derives from the four functions set out by Henry Oldenburg for the *Philosophical Transactions* in the seventeenth century and a fifth function enabled by the digital age[1]:

1. *Registration*. Dates (submission and publication) and author names are stamped on the article.
2. *Certification*. Quality control is ensured, usually through peer review.
3. *Dissemination*. The contents are intended for a specific scholarly audience, and readily accessible.
4. *Archiving*. A final version of each article is archived.
5. *Discovery*. The means by which the publisher enables scientists to find articles, text and data.

Journals achieve these through the use of a workflow applied and overseen by dedicated staff. Journals follow protocols at each point in the workflow, because they need to apply the same rules to each and every manuscript all the way from submission to the publication decision. Trained staff ensure every step of the workflow, and apply their expert judgment in decision making.

Journals live in challenging environments. Editorial offices need to cope with numerous routine issues and unexpected problems and still obtain timely, thoughtful and thorough input from busy editors and non-recompensed reviewers. Doing this efficiently depends on staffing, organization, commitment of the editorial office and board, and journal leadership.

> *Why journals are secretive.* There are two good reasons for why journals are secretive. First, they handle confidential information: authors', reviewers' and editors' identities, comments and reports. Second, they base their reputations on service: the time it takes to render a publication decision, the quality/expertise of peer reviews and editorial assessment, and the fairness of the decision (see Chapter 14). Journals do not want what are sometimes complex situations to be misinterpreted by outside observers. Journals may also be secretive for less laudable reasons. They can make questionable judgments and, knowingly or unknowingly, be biased (gender, geography, career stage…). Although journals may have internal checks and balances, they are rarely subject to independent inquiry.[2]

The Roles

Journals evaluate manuscripts and publish those that meet certain standards. Putting this into action requires experts in different roles.

Chief editor. The chief editor is responsible for journal decisions and in particular, publication decisions. The detailed responsibilities however vary from journal to journal. These can range anywhere from journal policy to following the progress of manuscripts from submission to decision, to only intervening in major events such as deciding whether to review a manuscript and making the publication decision. The chief editor is advised by members of the editorial board. Not all journals have a chief editor. Some are run by a managing editor or a committee of senior editors. Chief editors either have fixed tenures or renewable or open-ended terms. Some journals such as *Nature* and the *Trends* journals have professional chief editors, whereas most disciplinary journals are run by academic editors.

Editorial office. The editorial office is overseen by the managing editor and is the journal hub. It ensures smooth running of the workflow. The office notably liaises with the chief editor, the editorial board, reviewers, authors and the publisher. Editorial offices are increasingly managed by and within the publisher's office.

> *Workflows* are how an editorial office organizes its handling of manuscripts. The workflow determines who actually does what and when. An editorial office will have a basic workflow that addresses all of the routine events from manuscript submission to the publication decision. The workflow will be flexible enough to deal with unanticipated issues. A well-run office will be consistent in the handling of each manuscript; that is, obtaining the evaluations and applying standards in making publication decisions in a timely fashion.

Editorial board. The board is a group of dedicated, reputable scientists who contribute their time and expertise to assessing the soundness of manuscripts. They represent the subject areas covered by the journal. Some editorial boards have section editors or associate editors who take on additional responsibilities. Depending on how the journal operates, the number of editors can vary anywhere from a few to more than 1000 (in the case of "megajournals"). Journals with fewer editors tend to have lower manuscript submission rates and require greater time investments from each editor.

Handling editor or "editor" is the member of the board to whom a manuscript is assigned. The editor will have qualifications (subject area, technical expertise) permitting her to evaluate the manuscript. Editors may be responsible for one or more of the following: recommending whether a submission goes for external review, choosing reviewers, acting as a reviewer, arbitrating reviewer reports, providing comments for authors and a publication recommendation to the chief editor, and advising on appeals. Their service term may be fixed or open.

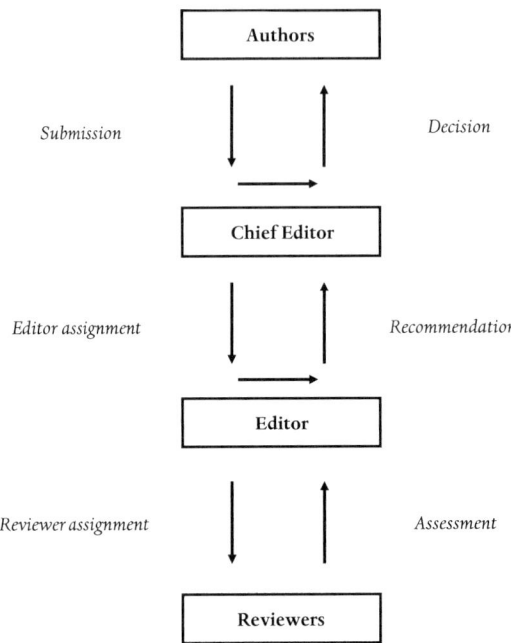

Figure 12.1 Simplified diagram of journal operations, only showing interactions involving authors, the chief editor, the handling editor and reviewers. The editorial office oversees all operations, interacting with each of these four players to ensure smooth, efficient completion of each task.

Reviewers. Reviewers provide independent assessments of a manuscript. External peer review both extends the expertise of board members and reduces potential in-house bias. Reviewers are generally scientists at large, but some journals have an extended, dedicated editorial board that reviews manuscripts. Good reviewers (quality, on time reports, and reviewing frequently) are sometimes invited to join editorial boards.

Authors are the source of a journal's science. They usually choose the journal rather than the reverse, though some journals have dedicated editors who actively solicit author submissions. Journal choice is influenced by author needs (speed, likelihood of acceptance) and journal characteristics (reputation, impact factor, readership). Most journals do not differentiate the way a manuscript is handled based on whether authors are members of an academic society, editorial board members or scientists at large.

Readers are not only influenced by published content and transmit this through citations, but they are also potential future authors. "Reading" extends beyond literally reading articles, and includes text mining, extracting data and finding methods.

Academic societies. Some journals are overseen by an academic society that interfaces with the editorial office and editorial board, some managing all processes themselves (operating as a full publishing service) and others contracting with a dedicated publisher. Some academic societies have a role in determining editorial policies and the selection of editorial board members. Members of the society may have reduced costs when publishing in the society's journal.

Publisher. The publisher's responsibilities include processing accepted manuscripts including assigning a DOI,[3] copyediting, proofreading, typesetting the manuscripts, managing the journal website and publicity at meetings and conferences, hosting pdfs and diverse metadata, printing and distributing paper copies of the journal and financing journal operations. Some publishers oversee journal operations including hiring the chief editor. Publishers are typically either a private commercial company specializing in academic publishing, a university press or an academic society. The former is often referred to as "for-profit," whereas the latter two are often called "not-for-profit". All three types must have revenue that exceeds costs in order to survive and prosper, and while this is referred to as "profit" for companies, it is called "surplus" for university presses and societies.

Gender bias. Helmer et al.[4] used public information about author, reviewer and editor identities from 142 *Frontiers* journals in the biological and physical sciences, engineering, health, the humanities and social sciences. When correcting for possible bias in their analysis, they found significantly lower than expected numbers of women as reviewers and authors and the same trend for editors. There was considerable disparity among journals, with minima of about 10–15% women and maxima of 30–50% women in different roles. There are increasing trends in the representation of women over the period 2007 to 2015, and significant effects of same-gender preference, for example, female editors solicit more female reviewers than do male editors, and vice versa. One of the main issues not accounted for in this and many other studies is that women are still leaving science in large numbers— in part because of the hypercompetitive culture—despite moves to address imbalances. This makes the appearance of trends in gender equality seem overly optimistic.

What Authors and Journals Want

Journals and authors are both after impact. An impactful paper is perceived as interesting and important, and ultimately changes readers and science. Journals benefit from impact both in their image and through attracting more, potentially impactful submissions. Authors gain in their reputations and authority. In striving for impact, both journals and authors contribute to increasing knowledge and understanding.

Despite these similarities, there are subtle differences in what authors want in a journal experience, and what journals want from authors.

The author wants her manuscript to be published in the choice venue based on merit. "Choice" can mean many different things. An author is likely to try prestigious, high-ranking, reputable journals first, knowing that even getting peer reviewed would be an accomplishment. Should her paper be reviewed, she is prepared to conduct the revisions necessary. But, like many, she does not view revising as an enjoyable experience and so does what is required and reasonable to satisfy the opinions of reviewers and editors.

The journal wants to publish choice science. Each journal will have its own criteria of what "choice" science is, and they typically include one or more of: importance, novelty, topicality and scientific quality. Identifying "choice science" marshals the contributions of the many players in the previous section. Chief editors cannot possibly give the same amount of attention to all submissions, and therefore only spend significant time on papers that have a reasonably good chance of acceptance. To achieve this, journals put a premium on identifying the most likely papers to survive peer review. Those that do not make the cut are desk rejected.

Should the journal decide to have a paper peer reviewed, then they solicit external experts. Reviewer assessments will help the editor decide whether publication is still an option and, if so, the conditions necessary via a revision. Importantly—and in contrast to many authors' own inclinations—the journal wants authors to *amply* satisfy reviewers and bring the study up to its fullest potential. The journal will encourage certain revisions and require others as a condition for further consideration or acceptance.[5] Journals are thus in a position of power. The ideal experience for a journal is that reviewers identify each and every shortcoming and the authors address them all successfully and to the fullest extent.

Journals therefore search for more than what many authors are inclined to give. In the extreme, summary peer reviews that result in a manuscript being accepted with minor revisions may delight certain authors, but actually signal a failure in peer review. Like the many forms of reviewer bias, summary reviews do a disserve to authors, journals and to science.

Journal Selectivity

"Why did the journal reject my paper despite globally positive reviews?" Of course, there may be many reasons specific to a particular manuscript, but all else being equal, rejecting a good paper suggests that your study was competing with many other good papers and possibly some great papers. The journal was selective.

There are two basic models in the way journals select papers to publish.

The first are journals publishing all papers meeting *acceptable* scientific standards—so-called "soundness but not significance."[6] Notable examples are "megajournals" such as *PLoS ONE* and *Scientific Reports*. These journals are of the philosophy that science deserves to be published regardless of its novelty or importance, or whether results such as negative, positive or inconclusive, as long as the paper is in scope, the conclusions are supported by the data, and the article meets certain technical and ethical standards of reporting. Because such journals have fewer publication constraints, they tend to have high acceptance rates (typically greater than 50%), and because of these high rates, can apply economies of scale (since a higher percentage of submitted

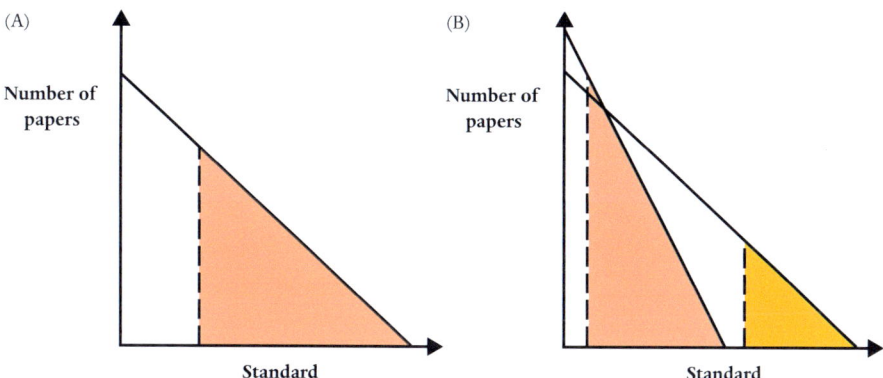

Figure 12.2 *Journal selectivity.* Shaded areas beyond the vertical dashed lines show the manuscripts accepted for a given journal. (A) Journals that accept all manuscripts beyond a minimal standard. (B) Two journals with different levels of selectivity. The orange journal receives more higher standard (perceived importance, interest, scientific quality) submissions and can be more selective than the red journal.

papers contribute to costs) and maintain lower publication costs compared with top ranking Open Access journals.[7]

The second—and vast majority—are journals that introduce added selectivity into their decision-making process. Selectivity may be in one or more of: scientific quality, novelty or importance. In making their decisions, journals will use some combination of a ranking system and the editor's views. Other journals will add to this a limit in the number of articles published per issue, meaning that if submissions increase/decrease, rejection rates also change. Journals with increasing submission rates can increase their selectivity—and as a result, likely their impact factor as well—by simply keeping the number of accepted articles constant over time.

> *Journals can make mistakes.* Just because a journal is selective and of high rank does not make it immune to error. These can range from publishing papers of low standard in statistical design and reporting, to those that are fabricated or with fatal flaws and are subsequently retracted.[8]

The Challenge of Peer Review

Many of the above elements are unsurprising to scientists with some publication experience. You submit a manuscript to a journal, it either comes back desk rejected in days or perhaps a week, or goes out for review and you hear back from the editor, anywhere from 1–6 months later. If you have submitted many manuscripts over your career, then you know that occasionally something will go wrong, and you not only have had an unpleasant experience, but a good story as well. What you probably will not know are the underlying reasons.

Well-run journals are not immune from contentious publication decisions. One of the main culprits is not having enough expert information to make a decision that meets journal standards and author expectations. Usually, journals require two to three reviews, but some manuscripts—either because of their complexity or limitations in the scope or depth of reviewers—might require several or even more. There are however norms in peer review, and most editors are reluctant to go beyond four. Thus, a vigilant editor will solicit three reviewers with the desired expertise, and eventually get a fourth should she feel that there are major gaps. This highlights how options for additional reports are open early in the review process, but become few or disappear as time elapses and the decision approaches.[9] The handling editor is key in such situations, since it is she whom the chief editor can rely on to provide her own comments on the manuscript and, in cases of highly divergent reviews,[10] arbitrate in making her recommendation. It is not unusual that the editor's views will diverge from those of one or more reviewers, and even with the review consensus. A frequent example is an editor who recommends major revisions, despite one or more reviewers suggesting rejection. In such instances the editor will provide explicit reasons to the chief editor and to the authors for disagreeing with reviewers.

Even the most professionally run journals can disappoint. Obviously, no author enjoys having their manuscript rejected. And even situations where a manuscript is revised through one or more rounds and finally accepted can be perceived as a negative experience, particularly if the required revisions were arduous and not (in the opinion of the authors) necessary.

To drive the above points home, below I present several procedural factors that can lead authors to believe that their manuscript was mishandled or the journal unfair in their decision.

Less appropriate reviewers

When a submission is to be sent for external review, the handling editor examines the manuscript and references therein, reviewer names suggested by the authors, her own reviewer choices and names suggested by the manuscript processing database. The editor then establishes an ordered list of reviewers based on a number of criteria, including no apparent conflicts of interest, appropriate scientific background, and quality and timeliness of the reviewer's eventual previous assessments for the journal. The editor is therefore *very much aware* that *some* reviewers on the final list may be unknowns in terms of quality and reliability. She also cannot be sure whether previously flawless reviewers will be up to standard for the current review. Most importantly, the editor cannot oblige the preferred reviewers at the top of her list to accept the assignment. This means that some of the actual reports received may be from reviewers who are less experienced, somewhat outside of the manuscript's main theme or less dependable in terms of time invested in the report.

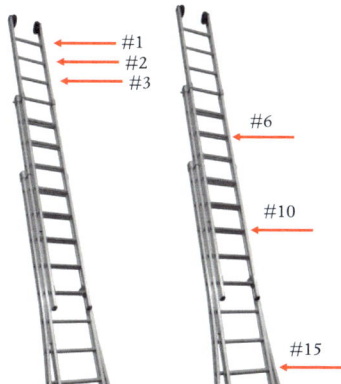

Figure 12.3 Authors might imagine that the reviews received were the first three invited (left), whereas in reality it can take many invites to get the required number of reports (right).

Late, rushed reports

Busyness is a problem for peer review, both because the star reviewers are often over-solicited and since those who accept to review can have unexpected demands on their time, meaning that their reports are delayed, never submitted or submitted but of low quality.

A single late report can delay the publication decision, sometimes by months. A well-functioning editorial office will avoid soliciting reviewers who are habitually tardy and/or provide low-quality or summary reports. However, some reviewers who have good past records may be late with a report, and the editorial office will either wait until the report is submitted or, should multiple reminders go without a response, notify the reviewer that her report can no longer be received. In the latter case, the journal may solicit a replacement report, either from a member of the editorial board or an external reviewer who has expertise in the subject area and can be counted upon to provide a rapid assessment.

> *Diversity promotes equity.* Murray et al.[11] recently analyzed the outcomes of peer review from thousands of submissions to the biosciences journal *eLife*. Among their important findings were that women and authors from outside North America and Europe were both underrepresented as reviewers, and that mixed-gender and international teams led to more equitable (less homophilic) peer review outcomes.

Lackluster reports

Lackluster reports (which are often also late) are equally challenging for an editor since, once submitted, they cannot be censured. Similar to reports from reviewers who are not experts in the area, an editor's only recourse is to carefully arbitrate the report. Despite the best intentions of and work by the editor, a rejected manuscript with one or more lackluster reports can understandably result in authors putting the blame on the journal.

Opinions

Reviewers evidently have opinions and these rightfully enter into their evaluations. Opinions and suggestions can be useful for authors in their revisions and for editors in making publication decisions, but sometimes opinions are too extreme or demanding, and are given too much weight by editors. "I strongly believe…the authors should…" can be interpreted by authors as requiring what may be unnecessary revisions. "I suggest rejecting this manuscript" without a supportive basis can lead the unwary editor to recommending rejection to the chief editor. In contrast, "I strongly suggest accepting this paper…" appearing in the comments to authors will likely rile authors should their paper be rejected, even if the reasons provided for rejection are sound.

Editors are responsible for differentiating critiques of fact (flaws in experimental design, analysis, incorrect citations) from opinions and suggestions in reviewer reports. The former tends to have more weight both in the publication decision and necessary revisions. Sometimes editors will not comment on reviewers' opinions and suggestions, and rather leave it to the authors to react and revise.

These points illustrate the challenges that editors face in ensuring decision standards and therefore fairness for authors. The question of who really has to be convinced for your manuscript to be accepted at a journal and how this may affect your writing strategy is addressed in the next chapter.

13
Who Really Decides?

The roles of editors, reviewers and authors in the publication process are loosely analogous to a court of law. Authors bring their case in the form of a manuscript to the journal (the court) for publication consideration. They present arguments in the cover letter for why the journal should take a positive view on publication. The chief editor functions as the judge, examining evidence provided by the authors and critiques/recommendations by external reviewers (the jury) and syntheses/recommendations by a member of the editorial board (trial counsel). A crucial step in the publication decision is approbation from the reviewers. This chapter discusses these analogies and the importance of writing a manuscript with reviewers in mind.

The previous chapter discussed how journals function and what they want. This chapter addresses the question of how they get what they seek and what this means for you, the author.

Journals Select Science

Recall that journals function by minimizing effort on papers they are unlikely to publish—through desk rejection—and focusing attention on those papers that have a good chance at surviving peer review. Journals make their selection on the latter by obtaining external reviews, and once received—for those papers that still look promising—requiring authors to revise as a condition for acceptance.

> *A decision is only as good as the deciding committee.* The diverse expertise and opinions of reviewers present a considerable challenge for journals to make "fair" decisions. To see the problem, consider the following illustration. We have a group of 1000 experts. We ask them each to individually evaluate and rank 10 manuscripts from the most to least publishable. No ties are allowed and the experts are not allowed to interact. Now imagine two scenarios. In the first, the experts are apportioned randomly into four groups of 250, each group of which establishes a mean ranking. In the second, they are in 250 groups of four. We select one group at random from each of the two cases. Any group from the former case will be representative of the remaining three, whereas a random pick from the latter case could very well hit the mean expectation on the nose, or be way off.[1] Thus, journals—in soliciting only two to four reviews—have to likely deal with limited information and sampling effects.

The chief editor interprets information and advice in making her publication decision. The information and advice come from only two parties: the editor handing the manuscript and external referees. Indeed, something that even many experienced scientists don't realize is just how much *each* individual expert potentially contributes to what will become the final publication decision. A chief editor decides whether or not to desk reject or send out for review, possibly based on a second opinion from a single handling editor. Should the paper be reviewed, only two to four people are involved, meaning that each report can have a big impact on the editor's recommendation and ultimately on the chief editor's decision.

The detailed interdependence of members of this small group turns out to be crucial in thinking about who you should really be writing for, as will be explained below.

The Court of Law

To better understand who is important in journal decisions, consider the following simile.

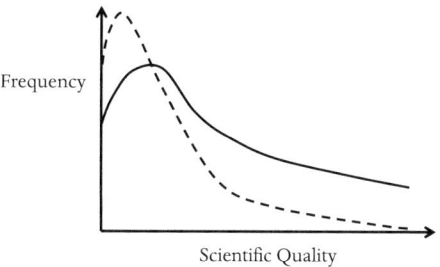

Figure 13.1 The centrality of peer review to selection and improvement of science. Hypothetical frequency of articles published in a journal in terms of scientific quality with (solid line) and without (dashed line) peer review.

Scientific journals operate in some respects like an inquisitorial court, since the applicant (authors) brings a case (submission of a manuscript for publication) to the attention of the court (the journal), and both the applicant and court stand to gain should the manuscript be correctly evaluated by the jury (reviewers) through fair interpretation of the evidence (the study) under applicable laws (selection criteria) and accepted by the judge (chief editor). The main contrast with courts of law is that applicants do not have full independent representation. The handling editor (counsel) comes close to this role and, despite being appointed by and associated with the journal, is expected to maintain some degree of independence. The handling editor is therefore the legal expert who weighs the evidence and presents a recommendation to the judge.

Who Really Decides

In the journal court of law, the function of the chief editor depends importantly on the quality of external and internal council. A chief editor of a journal receiving 1000 submissions a year will need to judge on average four new submissions a day, and make publication decisions based on peer review on perhaps two or three per day (since only a subset of submissions are sent for review). The editorial office ensures all of the logistics, but some of the more important problems arising will require the attention of the chief editor. As such, the chief editor deals with numerous different dossiers in any given day and needs the expert advice of editors, who in turn rely on the comments from reviewers.

Let's look at these two main checkpoints in more detail.

Checkpoint 1: Editors

Editors including the chief editor intervene at numerous junctures in the evaluation process. The first is whether or not the paper should be sent out for peer review. The chief editor decides based on the cover letter and the manuscript itself. As discussed later in Chapter 18 a great cover letter can be the deciding factor in sending a paper for review. But even when it is sent for review, the editor's reading of the manuscript itself will be determinant. A poorly written manuscript not only makes her job more difficult, but can muddle her perception of the science itself. Although a manuscript's communicative quality is not the principal criterion for peer review, if sufficiently poor, it can lead to the journal withdrawing the paper from consideration.[2]

Editors make rapid assessments when deciding whether to send a paper for peer review. If the chief editor is not sure, then she will seek advice from a specialized member of the editorial board. Editors will usually give a study some "benefit of the doubt," both since their expertise will probably only be a subset of that encompassed by a group of external reviewers and, importantly, because they cannot invest what may be several hours or more to thoroughly review each and every manuscript they receive (some journals do have "reviewing" editorial boards, but most rely on external expert assessments).

A manuscript that does not convince the editors therefore is a non-starter—unlikely to have survived peer review. If, however, the paper passes this first test, then the editors will want to evaluate more detailed arguments for or against publication. The paper is sent to external experts for in-depth peer review.

Checkpoint 2: Reviewers

Reviewers have two main duties. First, they comment on and critique the paper. Comments and critiques give authors a chance to improve their study, which is important both for possible publication in the target journal and for the quality of the study regardless of where the paper is finally accepted. Unless you have sought "friendly reviews" prior to submission (see below), peer reviewers are likely to be the only experts who provide thorough feedback.

The second way that reviewers contribute is to advise the editor on the study's publication potential in the target journal. Some journals do not explicitly request this information, but rather deduce it from the reviewer's comments. However, most journals do explicitly request that reviewers comment on—or rank—the manuscript's publication potential based on different criteria, including scientific quality, generality, importance and novelty. Journals sometimes also request what levels the study could attain if revised.

Each reviewer therefore contributes two interrelated types of information: comments and advice. Comments go to both authors and editor. Advice goes to the editor only, although some journals do transmit reviewer recommendations in the form of rankings to the authors. The handling editor will employ the reports in her own assessment and publication recommendation to the chief editor. Reviewers do not have the power to accept or reject manuscripts, but their assessments are the raw material for editorial decisions.

The bottom line is that you will need a consensus of favorable opinions from the editor and reviewers to get your paper published.

Write with Readers *and* Reviewers in Mind

Let's rephrase the essentials of publication decisions. To get published, editors need to agree to send your manuscript out for review, followed by reviewer reports that—very likely after revisions—contribute to convincing the editors to accept your paper. Editors poll reviewers because they are experts—they are readers at their most critical.

Writing a manuscript *with reviewers in mind* therefore addresses two challenges: getting your manuscript published and getting your published manuscript accepted by the readership.

Here are four non-mutually exclusive approaches to improving chances with reviewers.

The first, not surprisingly, is to try to put yourself in the place of the referee. Obviously, it helps if you have already had some experience evaluating scientific manuscripts. But with or without experience, think of what would constitute red flags in your paper to a critical scientist. The likely suspects are scientific shortcomings, flawed logic or unsupported claims, and unclear writing. Similarly, giving unbalanced treatment of issues or showing bias in citations is likely to be noticed by discerning reviewers—all the more so should the reviewer's own perspective be that which is critiqued, misrepresented or ignored. Evidently, an author cannot know for sure who will review the paper and there is no sense in doing a balancing act for particular individuals. Nevertheless, upholding the mantra of rigor is the surest course for both good science and positive reviews.

The second approach is asking one or more colleagues to act as a "friendly" reviewer. Even though colleagues may sidestep making potentially hurtful criticisms, they may be more objective than co-authors. Reviewing a manuscript takes time, and so it is important that your document is in good shape. Should the person agree to comment on your manuscript and do a conscientious job, then definitely communicate due thanks and mention their help in the manuscript's acknowledgments.

A third approach is to post your manuscript on a preprint repository. Interested readers may post or send comments, but systematic feedback via preprint servers is still uncommon. A complementary way to obtain comments is to solicit them through a specialized service such as *Peer Community In*... or *PreReview*.[5]

The final approach is to let a manuscript "sit." If you have had previous experience publishing you will have seen the positive impact of rest periods on the quality of your study. Writing furiously and continuously makes it difficult to see shortcomings objectively. Letting a manuscript sit enables you to mature ideas and to—metaphorically—see the study through a different lens. Sitting on a manuscript however has limits: what you gain in objective revision you may lose in time. The importance of "lost time" will depend on how you actually manage different phases of manuscript writing. For instance, if you are working with hard deadlines, then think carefully about how you divide up time among writing, revising and sitting. If the deadline is far into the future or you do not have a deadline, then clearly you have more leeway, but still you will want to limit the number and duration of inactive periods so that publication of your paper is not unduly delayed.

> *Reviewers have good memories.* You can "lose the trial" to get your manuscript published simply because one or more reviewers were unable to understand your study. Should you be fortunate and have a second chance (revisions or rejection with possibility to resubmit) and the chief editor solicits the same reviewers, then it's possible that they will be negatively predisposed to your resubmission, since they may have viewed the poorly prepared first submission as a waste of their time. This highlights the general point that (excepting double-blind review) reviewers see your name but not the reverse, meaning that a slipshod paper is likely to stick in their memories well beyond their review.

14
What to Expect from Journal Service

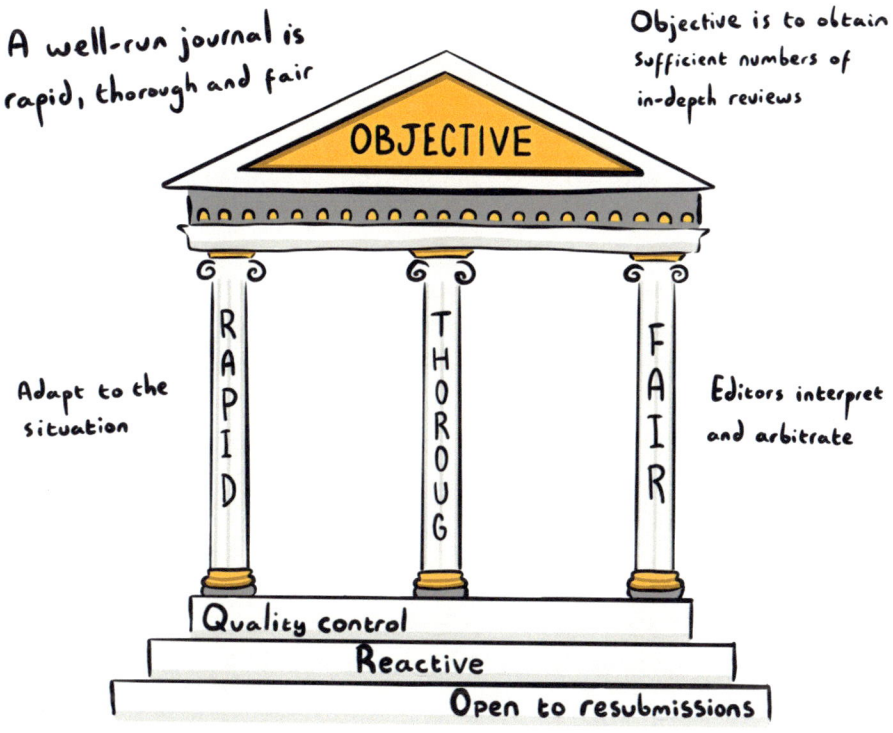

Journal service includes the manuscript handling system, notifications, the possibility of liaising with administrative, scientific and production contacts, but also less transparent features such as the number of reviews solicited, reviewer quality control, and possibilities for resubmission and for appeals. There is no charter for journal service. What should one expect? This chapter presents the essentials of professional manuscript handling.

A well-run journal has a honed and adaptable workflow. It would seem that all it needs to do is execute the workflow to select papers of choice at high professional standards. Although no journal can ensure that each and every manuscript will sail through their workflow smoothly, a well-run journal is less likely to encounter headwinds and more likely to be able to deal with issues before they become problems.

What are the hallmarks of a well-run journal? What goes on behind the scenes? What should authors expect?

A Bit of History

When I began as a young researcher in the mid-1980s manuscripts were prepared with a typewriter, photocopied and sent with a cover letter by snail mail to the journal. Typically, three or four copies were sent, one of which was for the editorial office, and two or three for prospective reviewers. Once received, the editorial office would send a postcard to the corresponding author acknowledging manuscript submission. Unless outside of journal scope, virtually all manuscripts were externally reviewed at disciplinary journals. Reviewers were not contacted beforehand: they were simply sent a copy of the manuscript with a letter explaining what was expected of them. Most reviewers complied, but typically took months to complete their reviews, and it could take a year or more to get a publication decision (although about 6–9 months was typical). The system was cumbersome in a number of ways, not the least of which was that the editorial office kept copies of all manuscripts and all correspondence.

> *Complex inner workings.* The editorial office at *Ecology Letters* operated via a combination of snail mail and file attachments from 1998 to 1999. We maintained paper copies of all manuscripts and correspondence for submissions up to 2001. A typical manuscript that went for two rounds of external review before being accepted had a file of around 50–70 separate documents. Flawless organization was required to manage and retrieve information on a daily basis, given up to 50 submissions under consideration at any given time.

With the advent of personal computers in the 1980s manuscript handling gradually began to change. Two big steps in the 1990s were email correspondence and electronic files. But it was not until the late 1990s and early 2000s that many journals switched to dedicated manuscript handling software. All documents and correspondence were now stored electronically on computers or servers, meaning that information was easily retrievable. Computer software systems, however, had their drawbacks, one being that many actions that could have been done automatically (e.g. reviewer invitation, reminders) had to be done manually. This was prone to error and very time-consuming.

Considerable progress in handling systems was made in the early 2000s. Within only a few years of the release of the first software, the first Internet-dedicated systems began to emerge. Internet-based systems are not only faster, better organized and more complete in the tasks they can accomplish, but also automatically execute repetitive time-consuming tasks. The time saved permits editorial staff to focus their efforts on guiding manuscripts, and more effectively deal with the problems encountered, through the editorial process.

Today's manuscript handling systems offer numerous possibilities for editorial offices, including reviewer databases, a range of modifiable letter templates, programmable reminders for tardy reviewers and a battery of journal performance reports.

The Editorial Office and Production

Nowadays, most journals have all the bells and whistles to handle manuscripts efficiently. When information is missing or invalid during the submission process you'll get error messages; should the initial online submission succeed, you may receive an email shortly afterwards stating that your manuscript has been "unsubmitted," because it does not meet all criteria for evaluation—this includes exceeding word or reference limits, having illegible figures, etc. Once any issues are resolved the submission will be validated online and/or via an email. The only other news that you might receive from the editorial office prior to the publication decision itself is whether your manuscript has been sent out for peer review.[1]

Production offices are typically distinct from the editorial office. Sometimes the editorial office will liaise with production on your behalf, but more often you will be contacted by an independent person who ensures a smooth transition from the accepted manuscript to the published article. This will involve transforming your accepted document into journal-formatted proofs and submitting them to you for eventual corrections and final approval. Sometimes, production introduces formatting changes or makes errors in transcribing your document, and so it's important to examine the proofs carefully. Certain displays (figures, photos) may not be sufficiently legible and, if this is the case, you should contact production to correct this. More difficult is to correct your own errors or awkward phrasings that were present in the accepted manuscript. Journals and production services frown upon this, both because manuscript editing should be complete at the time of final acceptance, and because there is a considerable gray area between correcting innocuous errors and either correcting important ones or introducing new material. As a guideline, any change that alters a result should be submitted to the chief editor for approval before making associated alterations to the proofs.

Production is not only not only copyediting[2] and typesetting. It also ensures that the article is discoverable through tagging with appropriate metadata. As data and other research outputs rise in prominence, production will also ensure that the article is digitally connected.

Times to First Decision and Publication

Times to editorial decisions and times from manuscript acceptance to publication *are* important, both for journals and for authors. By this I do not mean that journals should ever rush a decision or that production should rush final publication; rather, the very philosophy of "planting your flag" implies that the time and effort committed between manuscript submission and publication is in fact dedicated to the evaluation and publication processes.[3] Dead time—days, weeks or even months during which *nothing* happens—can be reduced with good editorial office organization, but can never be completely eliminated since an editorial office has no authority over tardy reviewers and limited influence on busy editorial board members. A well-run journal will anticipate delays before they occur and take measures to reduce them.

> *Ecology Letters.* Numerous delayed publication decisions in my own scientific career were the inspiration for *Ecology Letters* in 1998. *Ecology Letters'* mission was rapid publication. Making this work meant achieving a change in reviewer attitude, such that they would commit to submitting their assessments in no more than 3 weeks. This took several years to establish, and from 2002 until the end of my tenure in 2008, each and every publication decision was made within the prescribed times—depending on article type—of 4 to 8 weeks.

Many journals provide indications regarding their decision and publication times. This information may appear in editorials, included on the journal website, given in the submission-receipt email or indicated as actual times on each published manuscript. One should not put too much weight on these figures, since they might not differentiate desk rejections, revisions or rejections after peer review and accepted manuscripts. Moreover—with some journals—the reader has no way of knowing whether the published submission date is indeed the original submission date or that of a rejected and resubmitted version of the manuscript! Despite shortcomings, published indicators can be useful in journal choice (for more, see Chapter 16). They also provide a means for deciding whether or not to contact the editorial office should your manuscript appear to be out for review much longer than the journal indicated.

Number of Reviews

Peer review is the bread and butter of maintaining scientific and communicative quality. If norms were not as they are—and reviewers never burdened by reviewing—then editors would likely seek more than the standard two to four reviews per submission. In being limited to a small sample of reviewers, an editor searches for experts who cover different key aspects of the study, such as on the same or similar systems, on statistical analyses and on the central questions of the study.

It is increasingly difficult for editorial offices to get one or two, let alone three thorough reviews. Many journals seek a minimum of two but will go to three or four if they believe that the complexities of the study merit a more extensive assessment.

Another, practical reason for seeking several reviews is as "insurance", should some reviewers fail to come through or submit summary reports.

Difficulties in getting sufficient numbers of reviews increase the likelihoods of incomplete critique and a lack of consensus. Imagine a world where a manuscript received at least 20 reviews. Although two or more schools of thought might be represented and make consensus challenging, at least the handling editor can be confident that this is not a sampling effect! In contrast, in our world of few reviews, splits can be more difficult to interpret and sometimes important schools of thought just won't be sampled. Opposed positions require a handling editor to arbitrate a recommendation for the chief editor, as well as explain the journal's position to the authors and how this affects eventual revisions should the publication decision be not to reject the manuscript.

Building on these points, it is uncommon that in a group of three or four reviewers they will all have similar comments and make the same publication recommendation to the handling editor. Different reviewers have different backgrounds and perspectives. It's the handling editor's job to arbitrate such situations, as well as less contentious ones where all reports more or less coincide. Unfortunately, the editors at some journals do little or no arbitration, simply view each report as a "vote," and side with the majority. Authors of course are delighted when the majority position calls for revisions or to accept the manuscript, but if the opposite should occur and their paper is rejected—yet they believe that they can address the critiques—they are bound to be upset.

These issues highlight the complexity created by the norm of small numbers of reviews and the important role that handling editors play in arbitration.

Reviewer Quality Control

Editors are challenged to get thoughtful, deep reviews. Of course, they have access to massive reviewer databases and often recall the good reviewers and avoid poor ones. But the problem is not only about good or poor. The issue of obtaining high-standard reviews emerges perhaps counterintuitively because journals are free to invite whomever they should choose. Since journals do not communicate with each other, they tend to invite the same "good" reviewers repeatedly. Oversolicited reviewers will sometimes refuse to help out, or simply ignore reviewer email requests. This problem is discussed in more detail in Chapter 21.

In practical terms this means that some consenting reviewers may not have much reviewing experience or limited expertise in the subject area. With only two reports from non-preferred reviewers to go on, the handling editor may be challenged to reach the high scientific assessment standards expected by the journal.

Such situations can be addressed in the journal workflow. When the editor is concerned by the likely usefulness of reports, she has three main choices. First, she can get additional external reports or request that another member of the editorial board provides a review. Second, when making her recommendation, she can weigh in more than usual as an independent expert. Third, should the manuscript go back to the authors for revision, the editor can indicate that additional reports may be sought to come to a decision on the revised manuscript. Editors do their utmost to avoid this latter alternative that inevitably annoys authors, but for some journals it is considered necessary to make an informed publication decision.

Replying to Queries

Editorial offices are accustomed to receiving author queries. Some offices are very responsive and address problems professionally. Others unfortunately are unresponsive, are slow to respond or are not very helpful.

Author queries are usually one of the following:

1. Have you received our manuscript?
2. When will the publication decision be made?
3. I don't understand the reviewer's comment—can you please clarify it?
4. Can we please have an extension to resubmit the revised version of our manuscript?
5. How we can appeal the editor's decision to reject our manuscript?

Authors are sometimes uncertain regarding the appropriateness of contacting the journal with these and other queries. For example, authors might feel that they are bothering the editorial office by querying either of the first two, and that this might adversely affect the publication decision. Many authors don't even realize that they can send query 3 and, if they do, might not because it may take a lot of time to get a reply from the reviewer. For query 4, extensions are commonly granted, but they depend on details of the request. I discuss 5 in the next section.

Editorial offices evidently do not like being inundated with repetitious queries. This does not mean that queries will have a negative effect on the handling of your manuscript, but rather that they may not reply usefully to overly inquisitive authors.

Journals are usually responsive to questions pertaining to a publication decision, particularly when revisions are requested. It is good practice here—and more generally if you disagree with or are annoyed by reviewer reports or the publication decision—not to let your emotions enter into your correspondence. An editor may not reply to a strongly worded letter or will request that you reformulate your correspondence. Waiting a few days to let emotions settle before constructing a logical argument in a respectful tone should get you a response from the editor.

Soft Rejections and Appeals

Journals only have three main publication decisions: accept, revise, reject. "Accept" may come with a requirement for some minor revisions, and "revise" can span anything from very minor dotting i's and crossing t's to a major overhaul. "Reject," on the other hand, can be a "soft reject," where the journal either encourages—or leaves the door open for—a resubmission, or a "hard reject," which says that the journal will not consider your paper any further.

Soft rejections are a special type of major revision, where either it is unclear that revisions would sway the reviewers, or there are major flaws, and their resolution might have an effect on the study's findings and conclusions. A well-run journal uses the soft rejection to give authors the chance to address substantial concerns, and is usually not actually referred to by the journal as a "soft rejection," but rather worded in the following form: "although we are declining this version of your manuscript, we encourage you to revise and resubmit to our journal. Should you resubmit your study it will be handled as a new submission without prejudice." A soft rejection sends a message to the authors that they need to seriously revise to have their manuscript reconsidered by the journal and that eventual acceptance is not ensured.

Few editors can honestly say that they relish appeals. Appeals are made by usually upset authors who believe that the editors have made a mistake (also see Chapter 19). There are a number of reasons for appeals, broadly falling into two categories: either errors believed to have been made at one or more stages of the evaluation process, or, despite shortcomings in the manuscript, the authors believe that they should have been given the chance to revise and resubmit.

When handling an appeal, the editor will be looking for clear arguments that address *the actual basis for the rejection*. Generic examples of reasons for rejection include: fatal flaws in design or analysis, insufficiently rigorous science, erroneous inferences, or that reviewers do not view the study as sufficiently interesting or important. Whereas some of these could be appealed with cogent arguments, more subjective judgments (interest, importance) amount to opinions of authors vs. editor, and are usually not entertained further by the journal.

Limited recourse. An author disgruntled with journal service is unlikely to withdraw her manuscript, because the journal is "the journal of choice," withdrawing means starting all over with another journal, and because withdrawing is likely to irk both reviewers and editors. Should the paper be finally rejected, obviously the author might think twice before approaching the journal again. But should the paper be accepted—regardless of the experience—any publication costs will be due in full! Many authors never contact journals about what may be questionable incidents, possibly for fear that a complaint will influence decisions on any future submitted manuscripts. When they do query an incident, some journals will be responsive and take measures to improve their service, whereas others (arguably most) will do no more than send a simple apology.

15
Choices in Publishing

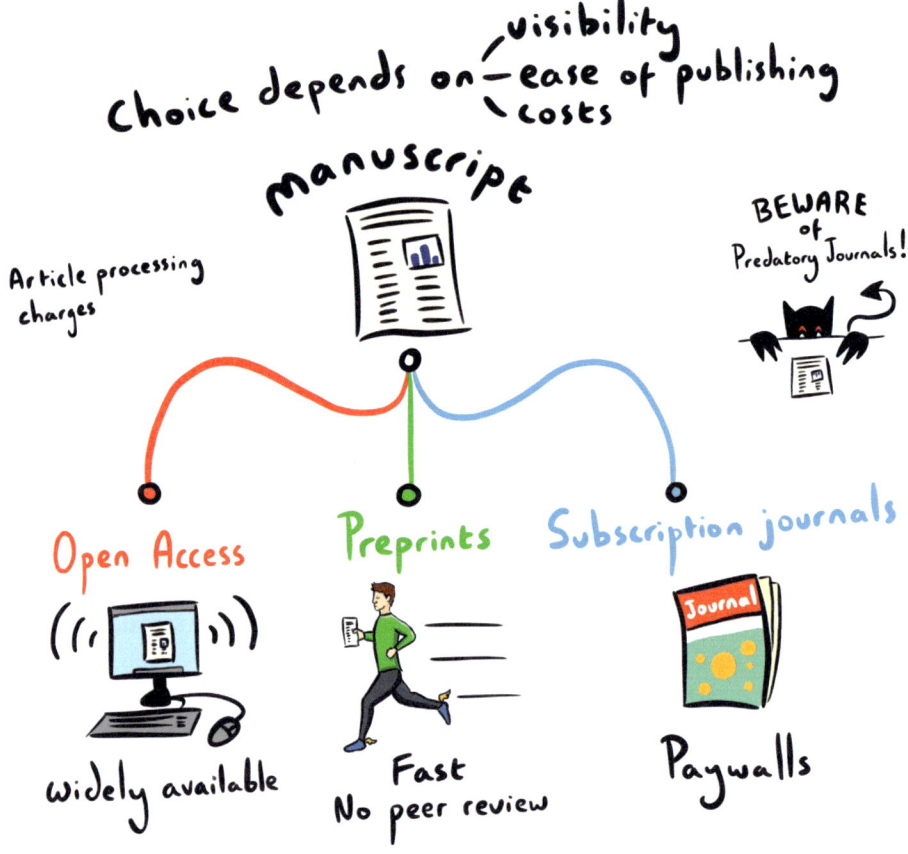

The world of publishing has changed considerably over the past three decades. Who can view published material, who pays for content and alternative models to classic peer reviewed journals are all important in considering where to publish your work. This chapter presents choices in publication venue and discusses pros and cons of each.

Navigating the world of publishing is increasingly challenging. Open Access, for-profit and not-for-profit journals, hybrid journals, article processing charges and impact factors, just to name a few of the terms you will encounter. Thirty years ago, the only real factor in publishing was journal choice itself. Journals were relatively few, and so choice was easy.

An Editor's Guide to Writing and Publishing Science. Michael Hochberg, Oxford University Press (2019).
© Michael Hochberg 2019. DOI: 10.1093/oso/9780198804789.001.0001

What was to become the "big bang" in publishing started in the 1990s with the emergence of personal computers. PCs had word processing, statistical and graphics software, and considerably facilitated manuscript preparation and submission. In the 2000s the big bang accelerated with the advent of the Internet. The Internet introduced email, journal websites, manuscript handling systems and social media—it revolutionized data storage, processing and retrieval.

Automation notably led to greater productivity per researcher and, along with increasing numbers of active scientists, generated a greater market both for journals and for publishers. Publishers had access to considerable quantities of data not only about science, but about where science is cited and how much governments, consortia, universities and individual scientists were willing to pay to publish and to read science.

The remarkable expansion in publishing models has created considerable opportunities for scientists, who can now choose where to submit their work based on the publishing economic model, conditions of manuscript evaluation, article access and eventual publication costs.

This chapter presents the features and shortcomings of different publishing models.

Some numbers. According to Crossref,[1] there are currently about 10,000 journal publishers producing over 60,000 journals. Active peer reviewed English language journals number over 33,000 and publish over 3 million articles annually. A select subset of these are indexed by Clarivate Analytics' Journal Citation Reports, which is composed of 2500 publishers, about 11,550 journal titles and 2.2 million new articles annually. Journal numbers were growing at an annual rate of 3% in 2001, increasing to about 5–6% in 2013. Over the period 2006 to 2016, the average annual growth in published articles was 3.9%.[2] All this for a total of about 70 million article records on the Web of Science Core collection.

Subscription-Only

Most journals generate revenue through subscriptions for access to published content. Subscription fees are either paid by consortia, research institutes, individuals or academic societies. Non-subscribers can also purchase individual articles. The main advantage of the subscription-only model for authors is that there is no Article Processing Charge (APC) (although many subscription-only journals do impose page charges or fees for illustrations). The main disadvantage is that some potential readers have no easy way to gain access to published material should they or their host institutes not be subscribed.

Open Access

Author and reader rights were innovated in 2000–2003 with the Budapest Agreement and Berlin Declaration[3] and the arrival of BioMed Central and then the Public Library of Science, initiatives to make published science freely available to everyone. Open Access (OA) additionally gives anyone the right to freely reuse the material in a publication (article, dissertation, book chapter, data) as long as the authors have published under liberal licensing terms (the authors retain copyright), and as long as the reuse credits the original authors.[4] Depending on the primary data source (e.g. Web of Science, Google Scholar, Scopus, etc.) and what content is considered as OA, current shares of OA articles vary between about 20 and 50%.[5]

The world of OA publishing is complex, but comes in two basic forms.[6]

In *Green OA* the published journal version is immediately available to subscribers only, but the authors (or publishers or librarians on their behalf) can post the final accepted version of the manuscript as OA at a repository and also distribute copies, but typically only after an embargo period. Once the embargo period is lifted, the copyright may or may not be held by the authors, and may or may not permit redistribution and reuse without permission.

In *Gold OA* the final version of the article is immediately available to everyone on the publishing journal's website and the author is free to post and distribute the document. Authors are generally responsible for any APCs. Authors generally cover this via their academic institution or a research grant. Assuming the author has signed a Creative Commons or very similar license, the article can also be posted in its entirety elsewhere by others or elements of it reused without permission as long as the authors are cited.

> *The Creative Commons* was established in 2005 as a non-profit organization to provide free legal and technical tools[7] to enable the sharing and reuse of knowledge. Numerous entities are affiliated with the Creative Commons, notably in academic circles, the Public Library of Science. The Creative Commons is most present on the Internet, but their mission is more general, promoting rightful ownership and information sharing.

Journals operate one of two main models using Gold OA. Some journals publish Gold OA articles only, and therefore in order to publish in these journals authors are responsible for any APC. In others—sometime referred to as "hybrid" journals—paid subscribers have access to all published articles, whereas authors can pay APCs to opt for their articles to be Gold OA.

Moreover, some subscription journals provide OA free of charge after an embargo period ("Delayed OA"), typically of 12 months. Delayed OA differs from Green OA in that the former is published on the journal website (similar to Gold OA),

Open Access Publishing

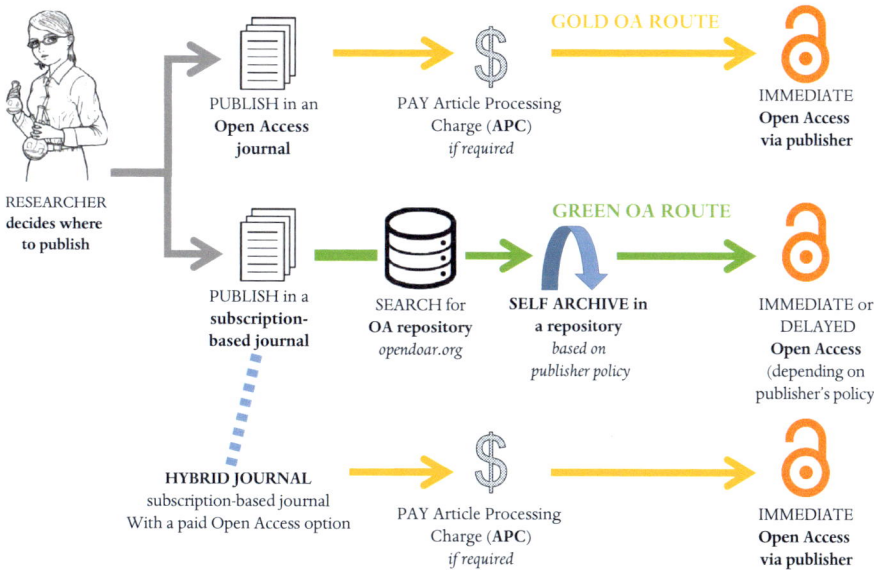

Figure 15.1 The basic routes to Open Access publishing (courtesy of Darren Chase).

whereas OA in the latter is posted at a dedicated repository by either authors or the journal.

The main advantage in Gold OA is that anyone can gain immediate access to the article and in some cases reproduce and reuse content without requiring permission as long as the source is cited appropriately.[8] The main disadvantage of Gold OA is that authors cannot publish in a preferred journal should they not have the financial resources to cover an eventual APC.[9] Many OA journals do have low or no APC; however, these tend to be very regional journals or in the lesser-funded humanities and social sciences, supported in-kind by institutions or other entities such as advertising and crowdfunding that provide alternative resources (so-called "Platinum" or "Diamond" OA), or *in extremis* may be "predatory" journals (see below). Some OA journals receive other forms of author payments, such as page charges, individual memberships and submission fees. Others—notably *eLife*—are supported in part by research funders, with the objective of promoting OA practices and limiting the financial burden on researchers and on the publisher. Cost models are discussed in detail in Chapter 22.

Preprints

Preprints are the freest form of OA, with immediate publication, minimal screening, minimal formatting, no costs, open feedback that can contribute to revisions should the authors choose and the freedom to subsequently publish (or not) in a peer reviewed journal. Preprints at certain providers have DOIs and so are citable.

The preprint norm is gradually becoming established across different disciplines. In the physical sciences, researchers regularly publish preprints on the arXiv, which dates back 30 years. Preprint servers have since emerged in chemistry (ChemRXiv), geosciences (EarthArXiv) and engineering (engrXiv), as well as the humanities (Humanities Commons), psychology (PsyArXiv), law (LawArXiv), social sciences (SocArXiv) and economics (RepEc). Biologists in particular have been very hesitant to go the preprint route until only very recently,[10] and are increasingly posting their manuscripts, particularly on the BioRxiv (Figure 15.2).

There are several interrelated reasons for why the preprint norm has taken some time to emerge in biology.

Lack of recognition. Posting a preprint on the Internet presupposes that the findings will be attributed to the authors. The paper is timestamped and receives a DOI. The preprint server shows the detail of how the paper should be cited. Nevertheless, readers who are unaware of the philosophy of preprints may simply regard them as not-to-be-cited progress reports waiting to be published elsewhere. Worse, upon seeing a preprint, why not rush one's own (similar) manuscript for formal publication in a peer reviewed journal? Or why not modify one's manuscript based on insights in a preprint...and not cite it? Any published article is of course susceptible to these same ethically questionable or unacceptable behaviors, but arguably less so than are preprints.

Stupid things inside. Although the argument can be made that authors who post preprints are careful about the quality of their manuscript because it is being made

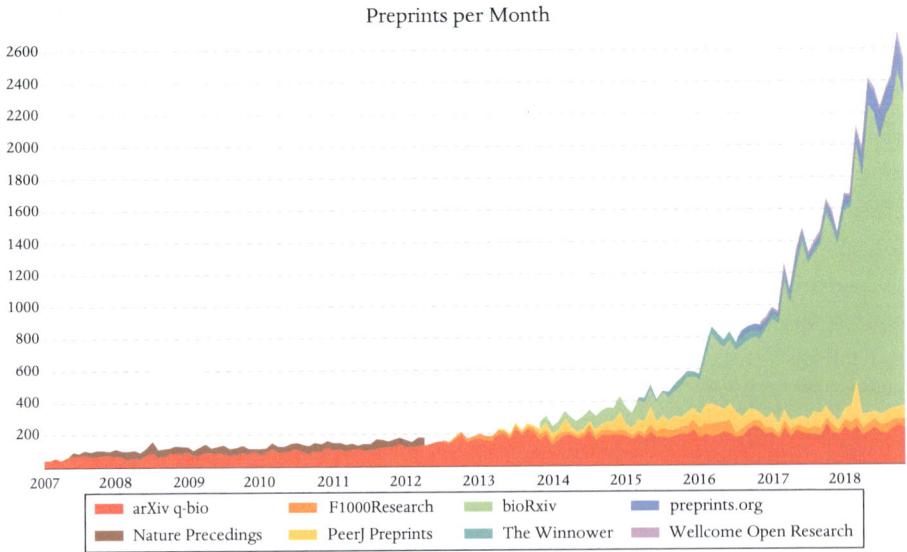

Figure 15.2 Biology preprint uploads per month for selected servers (figure courtesy of Jordan Anaya).

public, there is also the valid concern that authors may have missed (embarrassing) errors that would have been caught by reviewers of a traditional journal. Preprint servers do permit improved and corrected versions, but will having early versions seen, influence readers?

Dissipated motivation. Some preprints are never published in more definitive form, either because the authors didn't intend it, or because the manuscript was never accepted in a journal.[11] This introduces the complex issue of how depositing a preprint as a citable "publication" dissipates subsequent motivation to publish a definitive version.

> The *Ingelfinger Rule* was set out in 1969 by *The New England Journal of Medicine* to eliminate duplicate publication in the journal.[12] It was subsequently adopted throughout academic publishing to reduce confusion in attributing precedence, prevent saturation of the literature with redundant results and, together with media embargo, protect the uniqueness of the announcement for the benefit of the journal. The acceptance by journals of papers previously posted as preprints signals a relaxation in this rule.

Journals not publishing preprints. Most biological science journals refused to consider manuscripts posted on preprint servers until fairly recently. This is based on the foundational notion that a scientific study can only be published once—the "Ingelfinger Rule." Preprints, which may be posted in multiple versions, introduce the concrete problem of which version(s) if any to cite—especially as recognition of precedence.

Will Preprint Servers Replace Journals?

The gathering momentum of preprints in biology merits a discussion of their possible future. Will they eventually replace "classic" publishing?

The no-nonsense, rapid publication of preprints is the principal reason why an increasing number of scientists post them.[13] You can submit a manuscript with minimal formatting, it will be screened and appear online within hours or days. Unlike the vast majority of traditional journals, you can publish null or negative results. If you should so choose, you can submit to a preprint provider that has no peer review, or has it as an option, or requires reviews and author replies. Depending on the provider, reviews and replies may or may not be published online, and reviews may or may not be anonymous. There is usually no pressure to update your manuscript or even ever submit it to a more formal venue. Writing quality is not at a premium in a preprint. For the many scientists who are unable or increasingly less willing to bear the costs of publishing, preprint servers are a godsend.

It would seem that preprint servers offer what a growing number of scientists are looking for. Then why aren't they already replacing traditional journals?

The evaluation culture. Committees that make many of the important decisions in a scientist's career trust time-tested indicators such as the journal impact factor (see Chapter 23 for discussion of citation metrics). Preprint servers have no impact factor (since they are not covered by the Web of Science), citation rates of posted manuscripts are likely to be low since many scientists (at least in biology) are hesitant to cite preprints and—because they have no editors or peer review—they are not viewed as comparable with academic journals. Some funding agencies have recently changed their position, recognizing that ultimately science and its dissemination are important, and are now explicitly considering—if not encouraging—preprints.[14]

The scientific culture. There are also scientific reasons for why preprints are unlikely to replace traditional journals in the near future. Many authors post their manuscript *because* they have issues with peer review. But how is a reader to believe the veracity of a preprint if independent experts have not screened it? This transfers the onus of assessment to the reader. Am I going to spend my time reading preprints in detail when they could be faulty, and will be peer reviewed and published in the near future anyway? Moreover, would you ever consider providing detailed comments directly on the server? As opposed to formal peer review, is there any assurance that authors will heed your comments?

Preprint Initiatives

Despite some reticence, committees, scientists and journals are increasingly embracing preprint initiatives. Committees are under pressure to give greater weight to preprints in their assessments and decision making, and new initiatives such as ASAPbio seek to educate about issues in and the importance of preprints and peer review, with the goal of improving service and transparency.

Scientists too are contributing to bring preprints closer to the scientific norms of traditional journals. For example, Peer Community In…(PCI) assesses manuscripts free of charge as a validation service *before* eventually going to a more formal journal. The authors deposit their preprint on a server such as BioRxiv, arXiv, PeerJ or Zenodo. They then submit the same manuscript to PCI for peer review. After validation, the revised preprint is deposited onto the server carrying the PCI label of approval. But preprint review introduces the problem that journals will seek reviews again anyway. PCI is creating links with journals so that additional reviewers are not needed. PCI is also dialoguing with various institutions to get more recognition for validated preprints.[15] One possible outcome of the PCI model is that final, validated articles are published on preprint servers. Journals of the future would encompass peer reviewed preprint platforms.

Similarly, some journals aim to develop closer associations with preprint servers. For instance, the Public Library of Science and Cold Spring Harbor Research Labs seek to enable the immediate posting of submitted manuscripts to the major preprint server, the BioRxiv. Rather than go to a preprint server and *then* to a journal, the author does both simultaneously, and in so doing reduces unnecessary double reviewing. This evidently also limits the potential for preprint servers to become definitive publication venues.

A final example is an experiment by the journal *eLife*.[16] Their idea is that one of the reasons why peer review is becoming so unpopular is that reviewers sometimes have too much influence on publication decisions and required revisions. Editors should be doing much of what is delegated to external reviewers. Reviewers on the other hand should comment to improve manuscripts (see the courtroom analogy in Chapter 13), but it is the authors' responsibility to take these on board and make the changes to the paper that they see fit. In *eLife*'s experiment the initial screening by a handling editor is key, and a paper being judged acceptable for review is tantamount to acceptance for publication in the journal.

Predatory Journals

You likely receive on a regular basis unsolicited email invitations: invitations to conferences, to contribute to journals or books and invitations to join editorial boards. Some of these are credible, but many are not what they appear to be, and some are total scams.[17]

> *A sting operation.* Sorokowski and colleagues[18] created a fictitious scientist (Anna O. Szust) and used this person's name and fake credentials to apply to be on the editorial board of 360 different journals, 120 of which were from Journal Citation Reports (the most reputable journals), 120 from the Directory of Open Access Journals (a mix of journal reputations) and the remainder from Beall's List (suspected predators). Dr Szust was appointed editor in chief in four journals. Many journals did not reply or rejected the application, but in the end, 8 DOAJ and 40 predatory journals appointed her to their editorial boards.

The issue with predatory journals is that they lack the credentials of a reputable scientific journal. They prey on authors who are unable to publish their research in credible journals and/or have difficulties in covering moderate to high APCs. They also have relaxed peer review (if any), meaning greater ease and faster turnaround in publishing. Their use of the standard notion of "peer review" to attract submissions is therefore deceptive.

The parasitism of scientific publication is straightforward. There is *such high demand* for publication space that any semi-credible operation can dupe at least some eager scientists. The most vulnerable are early career researchers (some of whom are not aware of the telltale signs of a predatory journal) and scientists who are either unable to publish in reputable journals or get bonuses (money or promotions) should they publish in *any* journal.

The websites of these journals appear professional, invites look real and the journals have scientific names—and indeed model their names to closely resemble existing journals. Many publish few or no papers, probably because the scam in these cases is so obvious. Predatory journals charge an APC to authors for Gold OA, but the price is oddly low.[19] A good parasite causes little harm to its host.

Depending on criteria used, there are anywhere from several thousand up to 10,000 predatory journals,[20] and their numbers are growing.[21] A red flag should go up any time you receive an email from a publisher or a journal that appears to be legitimate, but upon further inspection, does not have the hallmarks of a credible journal.[22] As a rule of thumb, a journal that invites you to submit a paper without any specific reason is suspect.

- Be wary of any new journal without known editorial boards and links to known publishers.
- Investigate the journal's website and verify that the journal's head office has a credible address.
- See if the journal is covered by a recognized authority, such as the Web of Science or Scopus.
- Read the journal's recent publications for scientific quality.

Books and Special Issues

You may be invited to contribute a paper to an edited special issue of a journal or a chapter to a book. This is in recognition of your work, so good reason to be both honored and interested!

Publishing in an edited volume has its benefits and drawbacks. The main benefit is that you have more story freedom and less oversight from peer reviewers. Usually, an invitation to a special issue implies eventual acceptance of your contribution.

The writing freedom for an edited volume cannot be underestimated. An edited volume gives you a forum where you can develop your views in much more detail than in original research articles. It also lets you slightly lighten up the prose, making writing the paper a more enjoyable experience. Finally, an edited volume is likely to be consulted by genuinely interested readers, both specialists and scientists wanting to expand their horizons.

Edited volumes do have potential drawbacks. It is important that the co-editors have the ability to harness the best authors, professionally review and edit papers, and publish the issue in timely fashion. There is no way to be sure of the outcome of a special issue, particularly when it is an edited book, and bad surprises can happen, including your paper being rejected and publication delays. Good editing requires time, commitment, organization, diplomacy and, most importantly, communication skills. Good scientists are not always good editors.

16
Choosing a Journal

Imagine that you could fast-forward time and compare your paper's impact had it had been published in Journal A vs. Journal B. Publishing in Journal A may result a remarkable trajectory: many citations and the creation of a new lines of research. Publishing in B could be a positive accomplishment but lead to a more moderate impact or perhaps none at all. The same paper, two different journals, two different impacts. Indeed, venue *can* make a difference. This chapter will present different considerations in choosing a journal and how to develop a coherent strategy.

A Brief Conversation About Journal Choice

LABORATORY PI: "I believe that we should submit our paper to either *Nature* or *Science*. We are the first to demonstrate this new and important result that will influence research across disciplines. Our field studies involved more species and were conducted over a longer period than any previous. We integrated our field approach with mathematical models and showed a surprisingly good fit to the data."

POSTDOC LEADING THE STUDY: "I think it's a great idea to submit to a top journal. However, I'm concerned that we don't have enough evidence to firmly support our main conclusion! Really, I'm talking about the kind of evidence needed for the journals you mentioned."

LABORATORY PI: "These journals are certainly a long shot, but if our paper is rejected, then we'll just go to another high ranking journal."

This generic scenario is not at all uncommon. Strategically choosing a journal requires some experience, but it's not an exact science. PIs make mistakes. Sometimes they overshoot and select a journal that would greatly improve funding or career prospects. Other times, they choose a journal that corresponds more closely to likelihood of acceptance, but do not factor in the typical time required for a first publication decision to be made, and if positive, the likely additional time required for the paper to be accepted.

> *Most published papers are in first-intent journals, but most submissions are rejected.* Rejection rates of the top population biology journals are above 50 percent, and 80 percent rejection is not uncommon. But, when authors are successful, do they accurately target accepting journals? An interesting analysis by Calcagno and colleagues[1] indicates that by and large they do. These authors examined the percentage of articles published in a range of journals that were "first intents"; that is, the first journal approached. Surprisingly, 75 percent of published articles are, on average, first intents. This percentage decreases as expected with impact factor, but for the outliers *Science* and *Nature* they increase again to over 80 percent (this can be explained by these journals rarely receiving previously rejected papers from journals other than each other). As expected, higher impact journals receive more submissions, which, combined with the result that few papers are actually published by these journals, indicates that the 25 percent of papers that are not first intents are submitted to multiple journals before they are finally published, or possibly never published at all. A more recent study by Paine and Fox[2] not only supports these results, but suggests that these papers tend to fit the quality standards of the journal of final acceptance. The scientific standards and contribution of a study largely predict the range of journals in which it will be competitive and attain its potential.

Careful journal choice is important. Once your paper is published, it's for good. Although few could honestly say that they were in any way upset that their paper was accepted, some do come to regret journal choice afterwards. This latter point is actually quite subtle. Imagine that you choose journal X to submit your paper. It may have a higher impact factor than journal Y, but also have a somewhat different audience from most readers interested in your work. Your paper is atypical for X, but nevertheless accepted and published. Time goes by and you've had little feedback on your article and can't help but notice that it is rarely being cited. Although you could never know for sure, had your paper been published in journal Y, it might have had a larger, more receptive readership.

What criteria could you use in choosing a journal?

Journal Impact Factor or Rank

The journal impact factor (JIF) is updated yearly for journals listed in Journal Citation Reports (JCR). The JIF for calendar year Y is announced in year $Y + 1$, and is calculated as the number of citations received in calendar year Y for items published in years $Y - 1$ and $Y - 2$, divided by the total number of citable items published in these same 2 years. JCR produces ranked lists ordered by JIF, and because of sometimes magnitude differences in the typical JIFs between disciplines and subdisciplines, rank is often used to reflect impact. Thus, the top-ranked journal in one discipline may have a JIF = 30 and that in another a JIF = 3. See Chapter 23 for more discussion of citation metrics.

Consider the following scenario. Regardless of how you view the quality and importance of your paper, you decide to submit to the highest-ranked journal. You receive an email a few hours or days after submission with a desk rejection; if you are fortunate you may get some feedback from the editor. Let's say that you were prepared for this and you reformat your paper for, and submit to, the next highest-impact journal on your list. The same result—desk rejection—will probably happen again and again, until perhaps at the fifth or sixth try, where your paper goes out for review. You receive the reviewer reports and editor's decision a couple of months later to the effect that your manuscript was viewed as "interesting," but the editor did not think that the study was sufficiently competitive for further consideration in the journal. You are evidently disappointed, but you received useful feedback. You revise, reformat and possibly do some additional rewriting for the next journal in the loop.[3] Eventually your paper is revised for—and accepted at—a perfectly respectable journal.

Using rank or impact as the primary criterion for journal choice can be counterproductive in terms of time wasted, but also, perhaps surprisingly, in not meeting your objective of impact. This is because—paradoxically—your paper may get more reads and cites in a journal ranked 10th on the impact list than in any of the top nine. The JIF is the average value of all papers published over a given period. Impact factors have a number of issues in their interpretation, a notable one being that the variance around the average can be considerable and skewed.[4] This can mean that your paper could be in the lowest 10 percent of cited records for the top nine journals, and in the top 10 percent for the 10th ranked journal. For example, if your study involves both fundamental and applied aspects of disease and would be of particular interest to medics, then publishing in a higher-ranked fundamental science journal may lead to it being largely ignored by its readers, *and* missed by medics, who rarely read papers published in basic science journals. (This points to the underlying issue of scientists overemphasizing the use of journal titles in their decisions as to whether or not read a paper. See Chapter 24.)

Another notable problem in the use of impact factors for journal choice is how much of a difference *really* makes a difference? For example, would you choose journal X over journal Y simply because the former's impact factor is 5.3 and latter's is 4.9? Those who adhere to rankings would.

> *Impact and citation.* All else being equal, publishing the same paper in a high-impact venue will gain more attention than a low one.[5] Several factors contribute to this effect: higher prestige, a larger audience and more rigorous peer review (meaning that the published quality of the same study will be higher than at a low-impact journal). "All else" is not "always equal" however and so it is important to factor in the audiences typically reading papers in alternate journals.[6]

A Little Help from Your Friends

Building on some of the discussion above, an alternative way to choose a journal is to consider that colleagues are more likely to show interest in your paper if published in an "appropriate" journal. The appropriate journal may indeed be a high-impact, vitrine journal, or rather (more often) a highly respected interdisciplinary or disciplinary journal. You want to find the journal "where the shoe fits," and respected colleagues can provide a fresh perspective.

How can you put this journal choice strategy into practice?

Rather than automatically choosing to submit to a high-ranking journal, compare and contrast where you would like to see your paper published with *how collaborators or colleagues might react* to seeing your paper in that journal. There are two ways to do this. The first is to do a self-assessment test,[7] and then meet with your co-authors and ask: "If you were a scientist potentially interested in reading our paper, where would you react positively to seeing it published (and why), and where would you likely be perplexed by or negatively react to seeing it published (and why)?". The second approach is that rather than posing the above question to co-authors, you approach a few colleagues. If possible, try people with some publication experience and familiar with your research. Obviously, the colleague has to be willing to read your manuscript—even if only cursorily—to be able to come up with suggestions.

Decision Time (Especially for Students and Postdocs)

One would think that the time between submitting a manuscript and its publication is a crucial factor in choosing a journal. It is important, but perhaps less so than the time to the first publication decision. Let me explain.

If you are a PhD student or postdoc, you're always on a tight schedule and thinking about how you can meet research objectives before moving to another institute. What could possibly hold you back if you are producing results and contributing to writing papers in a timely fashion? Although busy collaborators can be challenging to work with, you can communicate with them and eventually move things forward. Good communication and organizational skills go a long way in your ability to transform research into articles (for more, see Chapters 25 and 26).

Things are different however with journals. Once your paper is submitted, there is *nothing you can do* to speed up the process. Some journals are very efficiently run and render decisions in days, weeks or at most a few months; others are ramshackle and may get back to you in 6 months or a year.

As discussed in Chapter 14, "official" times from submission to publication can be deceptive. Some journals systematically reject those papers requiring major revisions. This resets the official submission date to what is in fact the *re*submission date, but a reader will never see a trace of this on the published version. Such a journal appears to make rapid publication decisions! Times from submission to publication are also problematic since they may involve different numbers of revisions for different papers. Thus, the time from submission to publication is limited as a stand-alone indicator.

The time to the *first publication decision* is more revealing than the time to publication, because the former is the sole responsibility of the journal, whereas author-generated revision delays could enter into the latter. Importantly, once the first decision is known, you can reasonably predict the future time course of your manuscript. Thus, for example, should your paper be rejected, this is an indication that it will be a tough and long road ahead. Should your paper go into revision, the handling editor will likely give you a sense of the level of work involved, including whether your study will go out for another round of review.

The timeliness of a first decision is also important for two more subtle reasons. First, waiting for a publication decision can be nerve-racking, and long delays only to learn that a paper has been rejected can be disheartening for students. Some PIs make it a point to update students with any news about a submitted manuscript and—at least for "peace of mind"—will write to the journal's editorial office if a publication decision is late.

Second, timely first decisions mean that authors can more easily re-enter the mindset of their study and conduct revisions. Imagine the situation, which is not all that uncommon, of a first decision taking 6 months or more. The authors have not only moved on to other work, but have largely *forgotten* the scientific details of the submitted study! There can be considerable inertia in bringing everyone back up to speed and either conducting revisions or resubmitting the paper to another journal.

> *Fast rejection better than protracted acceptance. Ecology Letters* introduced the new norm of rapid manuscript assessment in ecology journals. Publication decisions were *all* made within several weeks of manuscript submission. Authors were evidently very satisfied with having their manuscripts assessed with little delay and published only months after the initial submission date. On several occasions the authors of rejected manuscripts wrote to say that, despite being disappointed by the publication decision, they appreciated not having lost very much time. One author even confided that he preferred being rejected rapidly to fighting for years to get a paper accepted. These and other similar observations drive home the importance of rapid publication decisions, especially for young authors.

Unfortunately, many journals do not publish first decision times with each article, and obviously one never sees decision times for rejected manuscripts. Other, sometimes more useful indicators include journals that publish editorials that report manuscript handling times, and discussions with group members and colleagues regarding their own experiences with particular journals.

What are reasonable first decision times? In the biological sciences, less than 2 months is good—more than 4 months should raise a flag. Something is definitely not functioning correctly with a journal that takes more than 6 months to make a publication decision, without at least notifying the authors beforehand of the reason for the delay.

Unless you are in a particular hurry, fast decision times should not be the primary reason to submit to a journal. Rather, *slow* decisions should figure among the reasons *not* to submit to a journal.

Reputation

Journal reputation has no metric. It integrates the quality of the editorial board, the JIF, the age and of course what people say about the journal. The latter is particularly important. Hearing consistently good (or bad) things about the handling of manuscripts and perceptions of scientific quality and interest (or lack thereof), are the most reliable ways to assess reputation. Reputation emerges from consensus.

Journals with great reputations got that way through years of quality service, positive word of mouth and consistently strong impact factors. Attaining high reputation requires considerable investment, thought and organization on the part of the editorial board, editorial office and any overseeing academic society or publisher. A great journal strives to consistently treat authors, reviewers and editors with respect.

Journals with good reputations receive greater numbers of good manuscripts. This means more competition for space, higher standards for getting published and higher impact. Attaining standards requires investment not only in the scientific study and manuscript preparation, but also in the quality of revisions. Less reputable journals tend to be less attentive regarding accepting marginal science and more permissive about revisions.[8]

Prices to Authors and Subscribers

Journal choice may involve elements totally unrelated to scientific reputation: either as a statement of support for a publishing philosophy and/or one's (in)ability to cover publishing fees.

Diverse parameters enter into the idea that certain journals or publishers are acceptable and should be supported, whereas others are not and are best avoided. Many scientists are bothered by the very concept that publishers make profits from their work. Publishing science is a market because of the high demand for space in reputable journals. The economics of publishing will be discussed in Chapter 22.

As much as OA is indisputably necessary to make all science available, the APCs often associated with OA create barriers for insufficiently funded scientists to publish in these journals. Part of the problem is that some journal APCs are determined by supply and demand economics, meaning that more attractive, higher-impact journals tend to impose higher APCs resulting in higher revenue for publishing companies.[9] Not all Gold OA journals are like this and indeed, many charge *no* APC.[10] The problem is that more reputable, international journals usually do, and this can enter into a prospective author's decision of where to publish.

Many cost-effective alternatives to high APC journals do exist, and there is evidence that authors are sensitive to price–quality relationships.[11] However, it is important not to confuse cost-effectiveness with low cost. Low-cost or free journals often carry with them other issues that a prospective author should seriously consider before approaching. For example, low- or no-cost Gold OA journals tend either to have limited readerships or be predatory. Preprint servers such as arXiv and bioRxiv do no manuscript handling outside of an initial screening and online posting. These can be valuable to rapidly distribute information and receive comments that will improve subsequent versions of the manuscript, but excepting initiatives such as Peer Community in…to review these manuscripts, they are usually unfinished versions with no external quality control. Finally, subscription-based journals are indeed often free of charge for authors, but as noted above, many scholars cannot gain access to published material, or have to pay one-time fees to do so. Among subscription-based journals, and a consideration for authors who want their papers read, is that academic society journals are often less expensive for subscribers than are non-academic society journals.[12]

History of Publication

Criteria such as journal rank, reputation or price, view publication as a one-off: any journal that fits your preferences is a potential home for your paper. This *à la carte* way of publishing ignores running themes that journals and contributing authors become known for. Thus, another way to view publishing is as a traceable, long-term process. You are building on a foundation by approaching a journal that is known for having contributed to the timeline. Journal editors recognize this and it does not hurt to mention the continuity of your submitted manuscript with previous work published (by you and/or others) in that journal. This of course does not prejudge whether the editor will eventually accept your paper, but she will certainly view your study as falling within the journal scope.

By publishing in a venue that has a reputation for science on a particular theme, you are contributing to the continued development of an intellectual compendium.

Mixed Strategy

One or more of the above criteria—rank, readership, reliability, reputation—will form your journal selection strategy. Your strategy will depend on your objectives, the information at hand and the opinions of your co-authors. PIs are likely to have their own views on the importance of different criteria, and may or may not open the choice up for debate, as in the conversation at the beginning of this chapter.

> *Reasons for journal choice.* CIBER conducted a multidisciplinary, international survey of scientific publishing by senior researchers.[13] Replies to the question "Reasons for choosing last journal" showed minor differences between top four replies (averages, where 5 = very important, 1 = not important): Journal reputation (4.50); Readership (4.21); Impact factor (4.04); Speed of publication (3.89).

PART IV
SUBMISSION AND DECISION

Your manuscript is written and the journal chosen. There are several important milestones that you will achieve on the road to publication. The first is that numerous people have contributed to the study in many different ways. Who merits authorship? When should this be decided? What is the importance of author order, how is it determined and when? Chapter 17 discusses the many issues surrounding authorship and good practice.

The second milestone is submitting the manuscript. This task is more than it seems—reputable journals often desk reject manuscripts within days of submission. Chapter 18 should convince you that carefully writing the cover letter to the editor will increase the chances that your paper will pass the first crucial step to publication: getting peer reviewed.

Peer review naturally leads to the third milestone—the publication decision. Chapter 19 presents different decisions, the importance of carefully reading the decision letter and—should you have the possibility of revising your paper—how to write the response letter to the editor and the reply to reviewer comments.

The fourth milestone is what happens to your data once your article is published. Chapter 20 describes the importance of data preservation and the rapidly changing world of data sharing as a central part of Open Science. "Open data" enables not only the verification so important for scientific progress, but also availability for often unanticipated future use.

17
Authorship

There is no objective method for determining authorship or author order. Common sense and journal guidelines indicate that authorship is merited for all those contributing "substantially" to one or more core criteria. Attributing authorship is therefore ultimately a PI's decision and relies on an honor system. Author order is also important, and typically follows one of a small number of norms. This chapter discusses the issues surrounding authorship and author order.

What Constitutes Authorship?

Authorship is the recognition of having contributed significantly to a study. Most scientists know significance when they see it. Person A wrote most of the manuscript,

Person B conducted all of the statistical analyses, etc. However, while a few contributors may have contributed to most of the work going into a study, there may be anywhere from one to hundreds of additional people who made far more minor contributions. Determining what and how much constitute authorship is one of the most difficult—and yet important—decisions in science. Unfortunately, there is no formula to determine authorship, and no dedicated authority that oversees authorship decisions.

Authorship is generally accorded or agreed upon based on the types and degrees of contribution. Contributions meriting authorship include: conceived and designed study, collected data, conducted analyses and wrote the manuscript. Although many journals are explicit about contributions that *do not* invoke authorship, some research groups nevertheless give authorship to individuals who have done any of these non-qualifying functions: obtained funding or provided facilities, participated in discussions or commented on manuscript, provided accessory unpublished data.

Authorship is based on a subjective assessment. Consider the following. Imagine that we could conduct and publish the same study twice. In conducting it for the second time, we decide to replace one of the contributors with someone else. Would this substitution have (from most to least significant): (i) Prevented the study from being completed? (ii) Prevented publication in the journal of choice? (iii) Had a significant negative impact on the quality of the study? (iv) Not changed the study in any significant way?

If the omitted person's only contribution was the sudden, brilliant insight at the origin of the study, then we would find ourselves in scenario (i). If the person was replaceable but ancillary to the actual study, then we would be in scenario (iv). This latter person worked full time and helped bring the project to completion. Both of these authors therefore made significant contributions, though one was indispensable (yet contributed only seconds) and the other replaceable (but spent months).

Real situations are often not clear-cut. For example, someone who contributes a few hours toward a non-specialized task that takes the team more than 100 hours to complete. Or someone who participated in discussions leading up to the study, and whose ideas contributed to the paper, but did not actually participate in writing the paper. Some PIs would view one or both as sufficient for authorship. Others would qualify them as meriting acknowledgment.

More generally, in some groups, certain types of contribution do not receive authorship and rather are acknowledged. These may include lab technicians, computer programmers, research administrators and students either volunteering or receiving credit for participation (but not engaged specifically for a thesis or dissertation). Other groups make it a policy to reward these and other contributions with authorship.

There is a huge difference between authorship and acknowledgment. Authors are the official contributors to the published study, and are recognized and credited as such by their peers, institutes, funding agencies, etc. In contrast, acknowledgment is official recognition *from the authors only*. Acknowledgments carry no other benefit beyond readers noticing who the authors thanked, and could not be included (without embarrassment) on a CV.

Misappropriations

The difficulties mentioned above aside, everything does not always go according to plan. People involved in a study sometimes contribute little of significance, either because they didn't complete what was expected, or they did, but it did not influence the written manuscript. This becomes an issue particularly if inclusion as authors was agreed upon at the beginning of the study.

> *Gifts, guests and ghosts.* Distortions in according authorship are probably common, but for obvious reasons difficult to determine. For example, some PIs attribute authorship honorifically to "gift authors," for what even the latter would regard as marginal or non-existent contributions. This may be done charitably, to appease or to gain favor. Contrast this with "guest authors" whose credentials are—in effect— used to make it more likely that the paper is accepted by a prestigious journal (and subsequently cited). Other forms of guest authorship stem from a co-author (on behalf of the guest) or the guest herself obtaining authorship based on a minimal contribution. Finally, and arguably the worst form of distortion is the intentional or accidental omission of deserving contributors, so-called "ghost authors."

Dealing with all of the eventualities in authorship decisions is not simple. Should the PI decide unilaterally without consulting team members? Meet with a few senior collaborators? Meet with all those potentially meriting authorship? One can imagine why the former and latter options can be problematic. Should the PI decide herself, she may overlook some contributions and unduly promote others. If she discusses as a group, then possible misalignments in her estimation of contributions and those of her team members will be exposed to everyone. This can be a healthy way to decide in small groups, but given the sensitive nature of the importance of authorship, may lead to challenges to PI credibility in larger groups. Regardless of which of approach is taken, sometimes either complex or strategic power relationships can be a major factor in authorship decisions.[1]

The absence of clear indication of authorship can be problematic, particularly for young scientists. Imagine working hard on a project, producing results and hearing about authorship only once the manuscript is ready to submit! Although this extreme is uncommon, it often happens that authorship discussions are delayed, either because the PI wants each team member to contribute to their fullest, or because the PI simply sees nothing wrong with discussing authorship at a late stage.

Justification and Responsibility

Journals want author credit to be given where merited. They typically have instructions for what constitutes authorship (and sometimes what does not) and may require a statement declaring the contributions of each stated author. The task of declaring "who did what" is the responsibility of the corresponding author (often the PI). The descriptions are usually general: "wrote the manuscript," "contributed to writing the manuscript," "contributed to revising the manuscript," and "agreed to the submission." The available choices may not reflect what you feel your contribution was. Some journals do provide space for more specific author write-ins, such as "conducted statistics and wrote the statistics section."

> *Contributor Role Taxonomy* or CRediT was proposed by Brand and colleagues[2] to "provide transparency in contributions to scholarly published work, to enable improved systems of attribution, credit, and accountability." This includes 14 different roles that qualify for formal authorship or inclusion (with explicit mention of the role) in the acknowledgments (see Table 17.1).

Table 17.1 CRediT – Contributor Roles Taxonomy (courtesy of Amy Brand; Ref. 2).

Term	Definition
Conceptualization	Ideas; formulation or evolution of overarching research goals and aims
Methodology	Development or design of methodology; creation of models
Software	Programming, software development; designing computer programs; implementation of the computer code and supporting algorithms; testing of existing code components
Validation	Verification, whether as a part of the activity or separate, of the overall replication/reproducibility of results/experiments and other research outputs
Formal analysis	Application of statistical, mathematical, computational, or other formal techniques to analyze or synthesize study data
Investigation	Conducting a research and investigation process, specifically performing the experiments, or data/evidence collection
Resources	Provision of study materials, reagents, materials, patients, laboratory samples, animals, instrumentation, computing resources, or other analysis tools
Data curation	Management activities to annotate (produce metadata), scrub data and maintain research data (including software code, where it is necessary for interpreting the data itself) for initial use and later reuse
Writing – original draft	Preparation, creation and/or presentation of the published work, specifically writing the initial draft (including substantive translation)
Writing – review & editing	Preparation, creation and/or presentation of the published work by those from the original research group, specifically critical review, commentary or revision – including pre- or post-publication stages
Visualization	Preparation, creation and/or presentation of the published work, specifically visualization/data presentation
Supervision	Oversight and leadership responsibility for the research activity planning and execution, including mentorship external to the core team
Project administration	Management and coordination responsibility for the research activity planning and execution
Funding acquisition	Acquisition of the financial support for the project leading to this publication.

Journals have no way to verify the actual contribution of each listed author. Rather, authorship is based on an honor system and journals apply this by asking for a declaration (usually a checked box during the submission process or a sentence that needs to appear in the submission cover letter) that all people meriting authorship figure as authors, and that only those meriting authorship are included as authors.

Authorship is not only recognition of contributions and due credit, it is also responsibility and accountability for the content of an article. The territory of accountability is first and foremost practical: if you conducted the statistical tests or collected the data, then should any of the results be called into question, you are the person (or among the persons) who is responsible for addressing the challenge and—should you choose to do so—replying. A formal reply typically involves all co-authors (who will co-author the published reply itself), and can vary anywhere from countering the critique, to accepting the error but limiting its importance, to (rarely) conceding and retracting the paper.

Author Order

Authorship is the official recognition of a contribution based on merit. But authorship is only part of a larger equation. Should there be more than one author on the paper—as is almost invariably the case—then the question of order naturally emerges.

Author order matters. It reflects the relative importance of contributions and influences how readers perceive this. Author order can be crucial for young scientists not only as an attestation of their real contribution (and pride), but also for their career prospects. If you are first author, you are indisputably one of the major contributors to the paper.[3] You would have done one or more of: conducting experiments, the analysis, or writing the manuscript. Your name will always appear whenever the paper is cited. But beyond a few key positions in the author list, there is no single accepted way to list author names.

The basic difficulty in deciding on author order is that the list is sequential, but contributions are multifaceted.[4] So, for example, if I wrote the whole paper and took one month to do it, how does this compare with a co-author having spent 2 months conducting all of the experiments? If I spent 1 month writing the Introduction, Methods and Results, and you wrote the Discussion in 3 months, where should our respective names appear on the author list?

Few co-authors mean fewer relative positions. These are usually the most straightforward situations to resolve, both because the real contributions of each person are likely to be known to all, and because it is easier to establish a personal rapport with a small number of colleagues than with a large group. Discussions can however become strained or turn to arbitration when contributions are about equal, when one or more co-authors view the publication as determinant to their future career, or when egos are involved.

The above situation mostly applies to two to several co-authors. Once the number gets beyond four or five, most of the positions have less meaning. There is virtually no difference between being second and third author on a four-author paper. And as the number of authors grows—perhaps counterintuitively—the importance of being last author grows too. A first author is still the lead on the paper, but when total author numbers get into the tens or even hundreds, then it becomes likely that the first author did less proportionally compared with a situation of only a few co-authors. In contrast, the senior author or PI—appearing last on the article—is perceived as the maestro: she gets major kudos since larger orchestras are more challenging to conduct.

So, what are the key positions?

Different disciplines have different norms and, as just mentioned, most recognize first authorship as significant. Until the 1990s, author names in biology were almost invariably listed in the strict (agreed) order of contribution. But with progressive increases in numbers of authors per paper, arbitrating relative positions has become futile. Contributions are increasingly specialized and incomparable, meaning that a linear order is no longer possible even on papers with only a few authors.

> *Authors need order.* Author order establishes different types and relative amounts of contribution, and can have important perceptual effects on readers and evaluation committees. Although typical author numbers vary from paper to paper and between disciplines, the average according to the Web of Science database is about five.[5] Ordering five authors appears easy…it's not. If PIs were to consider all possible orders, then they would have $5! = 5 \times 4 \times 3 \times 2 \times 1 = 120$. Therefore, PIs need to establish consistent rules for author order and be clear to co-authors about how the order was chosen.

In 2007, I co-authored a paper[6] arguing not only that authorship position is important in giving deserved credit, perception of contribution and implications for careers, but also that authors should seriously consider how they quantify Author Sequence and Credit (or ASC), and actually declare the method used rather than only stating who did what and positioning names.

We proposed four different methods that can be used to inform readers of what levels of credit *could* be given to each contributing author. The basic idea of ASC is that the impact of the journal or the impact of the paper itself can be apportioned to the different authors. For example, if there were only one author, and she gets X impact points based on the journal impact factor, then what is the logic that 100 co-authors should each get X impact points for publishing the same article? It could be argued that more authors result in a better paper, or that more authors mean more total energy or time invested. Nevertheless, the value added probably does not increase linearly with author number.

The four methods are:

Sequence determines credit or SDC. When contributions to the article decrease with each additional author, only the first author gets full credit, and a formula is applied to apportion reduced credit to each successive author. In the paper, we applied IF/n, where IF is the impact factor and n is the author position.

Equal contribution or EC. Here, the authors have all contributed approximately equally and will likely list their names alphabetically. When they declare "EC," they are saying IF/t, where t is the total number of authors on the paper. We suggested that a minimum contribution could be declared, such as 5 percent, which would mean if there were 20 or more authors, everyone gets at least $0.05 \times$ IF.

First–last author emphasis or FLAE. In FLAE, the first and last authors get more points than any authors in-between and the sum total in points is normalized to equal the impact factor. FLAE can be applied as is or modified to more complex situations where there is more than one first co-author or last senior authors.

Percent contribution indicated or PCI. This matches the requirement by most journals that authors declare their contributions in the published manuscript. Using PCI means that authors come to an agreement on how the IF is divided among them and declare this as percentages. This can obviously become unruly as author number grows, meaning that simplified demarcations would be more practical.

Using a system of credit such as those described above is sensible because the authors themselves declare their relative importance. A group could either declare the acronym in the acknowledgments of their paper, or when using PCI, indicate their percentage contributions. Other methods are certainly possible.

When to Inform Authors

I alluded above to issues in the notification of authorship. There is no all-purpose, easy way to do this. In some cases, early notification makes sense, whereas in others the PI will want to see that a contribution is being made before deciding. In yet others one or more persons who were not included as authors may approach the PI to query whether (or argue that) their contributions merit authorship.

While it is ethically sensible to inform collaborators of authorship sometime between a project's inception and engagement in manuscript writing, author order should be determined once all contributions have materialized, which is typically during the writing of the manuscript.

Fair policy. With the dual objectives of respect for group members and productive, enriching science, a PI should make authorship policy as clear as possible to new members shortly after they have joined the group or at the commencement of a research project. This includes what does and does not constitute authorship, and at what point in a project authorship and author order are determined. Should your PI not do this, consider discussing authorship policies with more experienced graduate students or postdocs.

One common mistake to avoid is determining authorship order early in cases where contributions to the study are not yet underway. For instance, if as PI you were to set out authorship order just when embarking on writing the manuscript, and have not yet discussed who will write what, then those authors who find themselves sandwiched between the lead and senior might very well contribute less than expected to writing, or worse, wait to see what others contribute before doing their share.

Deciding on author participation and order too early or too late can lead to problems. Once a manuscript is submitted, its authors and their order are fixed. Although very uncommon, the only changes that can then occur are during revision or in recasting a rejected manuscript for another journal. Either situation needs to be thought out carefully before taking any action, and should be first discussed with co-authors. In the former case, it is possible that manuscript revisions require new contributors (e.g. a statistical expert). This would need to be explained (including the positioning of the new author on the list) to current authors and justified to the editor, if resubmitted to the same journal. Very occasionally, a reviewer's report will change a manuscript fundamentally and the authors may decide to offer co-authorship to the reviewer.[7] If the reviewer is anonymous, then the authors can contact the editorial office to propose opening a discussion.

Avoid Disputes

There is nothing uglier than flagrant authorship misappropriation leading to a dispute. Gifts, guests and ghosts don't go unnoticed, particularly within research teams and host institutes. Collaborators who feel that they have been unfairly omitted—or included, but not in a fair position in the authorship order—may raise the issue with colleagues responsible for the paper. Such situations can be fairly dealt with before a paper is submitted. Once a paper is published however, the omission of deserving authors may result in article retraction by the journal.

18
The Cover Letter

Many view the cover letter as a nuisance that must be dealt with when a manuscript is submitted. Some journals make no mention of specific requirements for the letter; others not only require it, but also have specific guidelines. The cover letter is the opportunity to speak directly to the editor with the goal of convincing her to send your paper out for review. This chapter discusses the importance of the cover letter and its essential elements.

The cover letter is usually the last written task before submitting a manuscript. More experienced scientists may take the time to carefully write the letter before going online to submit; some however put it off to the last minute and largely improvise.

Journals often provide few if any guidelines for cover letters.[1] This leads authors to wonder whether the time invested really amounts to anything. For those who believe that the science carries the manuscript anyway, the cover letter is usually a minor inconvenience. After all, a desk rejected manuscript is unlikely to come back saying that the reason was an insufficient cover letter! Moreover, authors know that only a small minority of high-profile journals would actually not consider a manuscript based upon a summary cover letter. So why spend time on something that will never be part of a published study and about which many journals appear (or are) largely unconcerned?

The reason is that the cover letter is your direct line to the editor. It's the place where you can summarize why your study is timely, interesting, important and appropriate for the journal. The chief editor—and even the editor handling your manuscript—probably do not work closely to your area. It would take them some time and effort to understand the importance, relevance, novelty and scientific level of your work. They are busy people and there is a chance that—because they are hurried, your study is complex or the manuscript not well written—they do not fully grasp what your work achieves. The cover letter states in a no-nonsense way what is so exciting and important. It simplifies your story and ignores the rough edges.

The cover letter is important because it increases the chance that your manuscript will be reviewed.

Shows Motivation

Beyond being informative, the cover letter shows your motivation. It answers the questions of why the study was conducted, why you believe it's important and exciting, and, especially, why it's appropriate for the readership of this particular journal. A good cover letter suggests—without actually saying it—that you are prepared for multiple rounds of revision if that's what it takes to get your paper accepted. A good cover letter shows enthusiasm. It sparkles.

The goal of the letter is to excite and intrigue the editor so that she opens the doors to peer review. The cover letter will not influence the publication decision should your manuscript go out for review, but nevertheless, you need to succeed in the first step to even have a chance at the second, and so a good letter written in tens of minutes can turn out to be instrumental in whether your study is published.

Relatively Relaxed

Abstracts and cover letters have three things in common. They must both convey the essence of the paper, are usually about the same word length and you are likely to spend about the same amount of time writing each.

But this is where the similarities end.

The abstract is the scientific condensation of the study and never goes beyond what is contained in the manuscript itself. It is designed to get readers interested in reading further and is an *aide-mémoire* for scientists who have already seen the manuscript. Readers and peer reviewers, on the other hand, never see the cover letter. While the abstract is unlikely to influence the editorial process, the cover letter can.

A cover letter is your opportunity to talk to the editor in a relaxed way. Relaxed does not mean, however, overly informal, personal or sloppy.

A good cover letter:

- Is free of technical jargon and unnecessary detail.
- Can be read in just a few minutes.
- Highlights key arguments and information.

The cover letter is where you express why the scientific question is interesting, how your approach is an improvement over previous ones and what the results signify for future research. These elements will of course figure in the manuscript itself, but in a more technical way. The cover letter is the place to present your frank views while keeping to the facts. It can give elements not in the manuscript itself. For example: "During a pilot study, we observed an exciting, unexpected behavior, and decided to investigate its relevance to…In the present study we…".

Elements of Standard Cover Letter

Similar to the abstract, the body of the cover letter is a miniaturization of the manuscript. It will cover the background, puzzle, methods, results, conclusions and future directions. However, unlike the abstract, the cover letter stresses the highlights: the features of your study that will convince the editor of its quality, novelty and importance.

Although many manuscript submission interfaces now include rubrics where you can indicate any conflicts of interest, and preferred and non-preferred reviewers and editors, these choices should also be explained in the cover letter. While the chief editor is under no obligation to follow requests for or against particular editors or reviewers, most editors will respect such requests, especially when personal reasons are involved (which should *not* be described in any detail!). Editors are likely to be skeptical when more than a few names are listed.[2]

Cover letters are typically between half a page and a page in length, and should not go beyond two pages. They generally have the following elements (I give brief sample excerpts for each):

Salutation. "Dear Editor,"

Submit, title, journal name. "We are hereby submitting a manuscript entitled *title* by X, Y and Z, for publication consideration in *journal*."[3]

Norms. The letter should state that "all authors have agreed to and contributed significantly to the present submission," that "all people deserving authorship are included on the manuscript" and that "the manuscript is not under publication consideration elsewhere." Some journals have relegated these basic requirements to questions and boxes checked online during the submission process, but it's still good practice to include them in the cover letter. Moreover, many journals now accept that preprints of the submitted paper have been posted at a repository. It's good to mention this in the cover letter.

Lead in. This is where you introduce the editor to the science, for example: "A recent article by *Bonnie and Clyde* published in *the great Journal* argued that a central unresolved question in *this area* is to determine the role of *this factor* in the transformation of *this thing* to *that thing*."

Achievement. One or two sentences on why your study stands out. Use phrases like "first to our knowledge," "significant step," "exciting finding," "major advance," etc. It is important not to exaggerate: if you believe your study is an advance, but not major or earthshattering, then just call it "an advance." An editor will likely become wary should you oversell your study.

De-jargonized account of study. This is a user-friendly version of the abstract. I recommend including up to five cited references in the description and—if possible and relevant—include one or more published in the journal to which you are submitting.

Why this journal? Here you state your motivation for submitting to *this* particular journal. Recall from Chapter 16 that an important element in journal choice is its fit. Thus, arguments for "why this journal?" include: recent publications on the same question or debate, replying to a journal call to submit papers on a particular theme, matching the aims and scope of the journal (audience), and an important discovery for an influential journal.

Suggested/non-preferred editors and reviewers. "We suggest X or Y as appropriate editors, given their backgrounds in *this and that*." "We suggest that because of past conflicts, *name* not be solicited to review our manuscript."

Signoff. "We look forward to your publication decision, and please do not hesitate to contact us should you have any questions."

Salutation. "Yours sincerely,"

Some journals have specific guidelines for the cover letter and, although they likely coincide with elements in the above list, it is important to scrupulously follow any journal requirements or recommendations.

Submitting the Manuscript

Although there may be journal-to-journal differences in what is required for an online submission, most journals require the following:

- Author names, affiliations and contact details (and identification of the corresponding author(s)).
- Cover letter.
- Abstract.
- Keywords.
- Manuscript, possibly with figures, photos and tables in separate files; electronic appendices; any information not for peer review such as recent published manuscripts.
- Suggested or unpreferred editors.
- Suggested or unpreferred reviewers.
- Checked statements concerning authorship, data archiving, that the manuscript is not under consideration with other journals and that authors will cover publication charges (if any) should the manuscript be accepted.
- ORCID[4] identifiers. These serve to give a unique code to each individual author. This is useful since the same person may use different variants of their name, and because more than one person may share the same name.
- Granting agencies and possibly grant numbers.

Key to the submission process is listing preferred reviewers. Knowing that authors are unlikely to propose competitors or people they view as potentially critical, and are likely to suggest allies or like-minded scientists, editors treat the list as suggestions. An editor will consider these, add her own picks and establish a ranked list to invite.[5]

The accurate listing of affiliations and granting agencies is important, since it is an official declaration of their support: they get recognition credit once your article is published. Affiliations are not honorific: they include your actual employer(s) and other institutes where you were officially physically based and conducted significant work leading up to the final manuscript. It is good conduct to check with such institutes first if you are not sure about their policy for declaring affiliations. A similar approach is taken for declaring granting agencies that contributed to the research: some grants will be integral to the work and others peripheral, but nonetheless significant. Some granting agencies have specific rules regarding when acknowledgment is necessary and when it is forbidden.

Submitting a manuscript can be time-consuming and is not always completed in a single attempt. You can always push the "save" button and return at the later time. It's wise to keep backup copies of any extensive written material that you have entered into the on-line submission system.

At the end of the process, the journal will ask that you verify the html and/or pdf version of the manuscript. This is important both to make sure that any non-alphanumeric characters are reproduced correctly and to verify that any figures, photos and tables are correctly reproduced and legible to reviewers. Occasionally, one needs to adapt formats to produce a reviewable manuscript.[6] Once all is verified there will be a submission finalization step, and you should receive an email confirmation shortly thereafter.

19

The Publication Decision

The publication decision email is the moment of truth. Hundreds, perhaps thousands of work hours have now been judged and—in a single click—the decision appears. Our inclination to search for key terms such as "accept" or "reject" may blind us to important elements in the letter. This chapter explains why decision letters deserve careful reading and presents strategies for replying.

Some of the most exciting experiences you will have in your professional career are reading decision emails. Weeks or months have gone by and now a decision—based on the views of esteemed colleagues and journal standards—appears with a click.

Upon opening the decision letter, you will surely scan for keywords such as "accept," "revise" or "reject," and indeed the essentials of decision letters usually can be understood in a flash reading. But letters also contain important information—for example, the journal's openness to a resubmission, whereas a first quick reading suggested a firm rejection. Decision letters need to be read carefully.

Editorial Decisions

The decision letter should provide the following information:

- The publication decision.
- How the decision was reached.
- Guidelines for eventual revisions or resubmission.

Journals use a small palette of decision terminology. The basic distinctions are rejected, revise or accepted.

Rejection

As discussed in Chapter 14 there are two main types of rejection. A hard rejection means that (barring an appeal) the journal will not reconsider a revised resubmission. The wording varies but will typically use the term "reject" and either explicitly say that the journal will not reconsider the study, and/or suggest that the paper be submitted to an alternative venue. In a soft rejection, terms like "decline" or "reject," may be accompanied by a proposal to reconsider a revised study as a new submission. The editor may additionally use the term "without prejudice," which means that the decision itself to reject will not enter into consideration in evaluating any resubmission. When a resubmission is allowed or encouraged the editor will typically provide details regarding the concerns that need to be resolved and possibly details on how the manuscript will be handled (assessment by previous and/or new reviewers).

Revision

Many decision letters actually make no definitive decision, rather, the editor requests revisions before the paper can be considered further. The editor may qualify the overall level of revision as "major," "intermediate" or "minor." Conditions for the revision will usually be specified in the decision letter.

Revision is the usual route to acceptance. Editors may give indication of their interest in publishing your manuscript after revision. Seeing phrases or words such as "We are interested in considering….," "publishable" or "acceptable" are potentially good signs, and should be accompanied by specific conditions.

Editorial decisions to revise do not necessarily provide information about eventual acceptance. Authors sometimes understand "revision" to mean that—once revised—their paper is assured of acceptance. Unfortunately, this is not necessarily so. Even should you do your utmost to satisfy critiques, it is possible that reviewers are still not satisfied and the editor either gives you a second chance at revisions, or if sufficiently important concerns remain, decides to reject your manuscript. It is therefore best to assume that if there is no mention of "publication" or "acceptance," then the chief editor is not sure of whether a revised manuscript will gain sufficient support from reviewers, and a positive recommendation from the handling editor.

Acceptance

Typically, after one or more rounds of revision—should the editors be satisfied with your replies and changes made to the paper—you will receive a notification of acceptance and how to proceed to final publication. Sometimes a manuscript will be "accepted pending further revision," generally meaning very minor revisions.

Replying to a Decision Letter

You receive the decision letter and either it requests revisions or your paper has been rejected with the possibility of resubmission. In either case, along with your revised manuscript, you will need to provide a letter replying both to the editor and to the reviewers.

Writing a reply letter needs to be done with care—both in terms of scientific details and how you express them.

A reply typically has two main parts.

The first is a letter to the editor. The letter should bring the editor "back up to speed," since weeks or even months (in some cases of major revision) might have elapsed between the decision and your resubmission. The letter to the editor is where you restate the main advances of your study, focus on the positive parts of the assessments, clarify, explain or respond to one or a small number of important critiques, and eventually deal with any extenuating circumstances such as unreasonably critical reviews, clearly inappropriate (personal) reviewer comments and major changes to the paper that go beyond reviewers' comments. The latter three situations are serious and should be carefully considered before claiming (former two) or executing (latter one) them. The editor will examine cases of alleged bias or inappropriate comments closely and decide how to proceed. Should the claims be of a serious personal nature, she will likely consult the body (publisher or academic society) overseeing legal aspects of the journal.

Correcting or adding new results that were not mentioned in the original decision letter is unusual and must be reported and justified: at some journals this could be grounds either for rejection (should the editor call into question the scientific integrity of the authors) or requesting further review.

A typical letter to the editor will read something like:

> Thank you for giving us the opportunity to resubmit a manuscript entitled <Title> for publication consideration in <Journal name>. In this manuscript we tested the theory <Statement of theory> and found <Statement of main result>. This is the first

conclusive demonstration that.... The editor and reviewers, although expressing certain concerns, said that our study was "important." Reviewer 1 referred to our study as "a major contribution"; reviewer 2 said "this study is an interesting advance." We have taken the editor's and reviewers' comments into consideration in revising our manuscript. This has led to a greatly improved study....

The second part of the letter is the point-by-point replies to the reviewers' comments and the editor's (eventual) observations.

> *Editor indications are important.* Should you have the opportunity to revise, then look *carefully* for what the editor indicates requires particular attention. Obviously, you need to address *all* reviewer comments, but some may be more important than others to the subsequent publication decision.

Here are some general guidelines:

Be respectful. Remember that reviewers have altruistically contributed their time to improving your manuscript. There is nothing more annoying than sarcastic, aggressive or insulting author replies. Contesting reviewer comments should be done with due respect (see below).

Be firm. The flip side of the above rule is that you should be forthright in your replies. Say things the way you see them and explain why. Honesty and respect make for effective replies.

Be organized. Usually reviewers number their comments, but sometimes they don't, and sometimes it is even difficult to distinguish comments requiring a reply from simple observations. If the reviewer has not organized her comments as a numbered, point-by-point list, do your best to distinguish them by creating your own numbered list. Proceed by copying each comment (or an essential quote from it) from the reviewer's report and pasting it into your letter. Reply to each comment one by one with the heading "Reply."

For example, imagine that the comments from Reviewer 1 are unnumbered in a single paragraph:

Reviewer 1: "This, that the other thing, blah, blah, blah. Observation 1, 2, 3, blah, blah, blah."

Your point-by-point replies to this reviewer would be listed as:

1. This, that the other thing, blah, blah, blah.

    Reply: The reviewer points out, blah, blah, blah. We agree and modified our manuscript...

2. Observation 1, 2, 3, blah, blah, blah.

    Reply: The reviewer claims blah, blah, blah. We disagree that blah, blah, blah...

Request clarification if necessary. As per above, it is not uncommon that comments are unclearly written or difficult to interpret. In such cases and after consulting with your co-authors, either reply by first stating your interpretation of the comment, or if the comment appears important and you are at a loss for how to reply, contact the editor to ask for clarification.

Guide the reviewer. Most reviewers *want* to verify the authors' replies and revisions. But reviewers only have so much time and patience and therefore don't want to have to decipher ambiguous replies and bumble through the manuscript to locate the revisions. It is therefore good practice to begin a reply by recasting the comment. For example, "*Reply*: The reviewer questions our interpretation of the significant result in Figure 19.1. The reviewer proposes that the trend can be explained by… Our hypothesis to explain the effect is based on…". And, in addition to your reply to each comment, clearly explain how you (eventually) altered your manuscript. If the alterations were brief (one or two sentences), then state the alteration in your reply. It is also important to guide the reviewer by providing line and page numbers pointing to where the revisions can be found.

Details matter. All else being equal, detailed replies are more likely to sway reviewers than brief explanations. Make sure that your reply is complete and convincing without overwhelming the reviewer with useless details.

Beyond modifications to the manuscript, there is also nothing preventing you from providing information that will never enter the manuscript itself, or is only included in electronic appendices. An example of this is conducting new analyses. Your choice not to include what the reviewer may interpret as useful new information in the manuscript itself should be justified.

More generally, in writing your replies to the reviewers, be aware that—except in double-blind review and cases where reviewers sign their reports—the reviewer knows your identity but not the reverse. Your reputation is an added reason for why it is important in your replies to be professional, thorough and courteous.

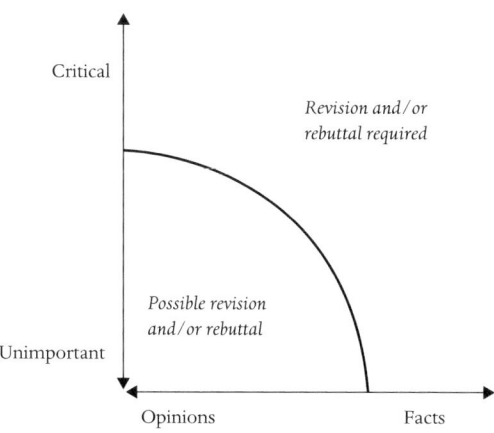

Figure 19.1 Hypothetical author responses to reviewers' comments. A revised submission should (i) reply to all comments in point-by-point responses and (ii) revise and/or provide clear rebuttals (or clarifications), particularly when reviewer comments involve critical facts.

Not Agreeing With or Not Understanding Comments

Editor and reviewer comments are intended to be constructive. But what if you do not agree with them or do not believe that a comment requires a change to the manuscript?

Don't agree. Should you not agree with a comment, first consider the possibility that the source is not scientific, but rather due to an ambiguity in the manuscript and/or the reviewer's misinterpretation. If stemming from ambiguity, you can reply: "We believe that the reviewer's contention <*contention explained*> stems from ambiguity in our wording on <*line numbers in manuscript*>. We intended to say <*explanation*> and thus agree with the reviewer's point. We have reworded these sentences accordingly." If from misunderstanding: "We believe that the reviewer misunderstood our explanation for the result on <*location in manuscript*>. Our result indicates <*explanation*>…". Otherwise, if the reviewer's comment is valid, but you disagree with it, then you will need to develop a reply that provides clear arguments and evidence supporting your position.

Don't change. Some comments need not involve changes to the manuscript. If you believe that a reply to the comment without revision suffices, then (barring obvious trivial comments) you should briefly explain why.

Appeals

Some authors do it systematically. Others should sometimes consider it, but don't bother.

Appeals are written because authors feel that their rejected manuscript has been misjudged. Some appeals vent anger, giving no coherent scientific reason for questioning the publication decision. Others are the authors "playing the lottery," not contending the reports in any detail, but rather summarily arguing their case and hoping for a chance to resubmit. A serious appeal contends the decision based on either a purported factual error in the assessments, or an error of judgment in the decision. Appeals were briefly discussed in Chapter 14 and I present them in more detail below.

A viable appeal presents solid arguments for why the *actual reasons for rejection* were either incorrect, or they were just, but rectifiable. Many potential appeals are non-starters because authors have either too little to go on, or too much to contest. For example, it's difficult to challenge an editor's (subjective) opinion that a study is not sufficiently interesting or important. It's your opinion against theirs. Rejections may also be based on major scientific shortcomings and the editor's view that even with revisions the study would not meet the journal's publication standards.

An appeal letter should begin by presenting the general context. This includes the formalities of the title, authors, decision date and handling editor if known. Then one or two sentences recounting the decision, the reasons given by the editor for the rejection and context of the appeal. For example, "Our study was rejected because it was judged scientifically incomplete, and specifically according to Reviewer 3 that it required additional experimentation. We are prepared to conduct this experiment, which we believe will satisfy the concerns of the Reviewer."

Following the basic proposal, if the appeal includes rectifying shortcomings in the manuscript, then give some detail as to how you would revise and why you believe this would satisfy the reviewers and the editor. You then want to "stack the cards" (if possible) in your favor, with verifiable, factual statements in the form of quotes, like "Reviewer 1 found our manuscript to be novel and important; Reviewer 2 said it was one of the finest studies she/he had ever seen." It is important that the appeal letter not list endless details; it should convince the editor of the basis of your contention and the proposed resolution. So, the basic structure would be:

- statement of context;
- basis of the appeal;
- stacking cards (if possible);
- some details about proposed revisions.

An appeal can express disappointment but should not be angry, personal or call into question the integrity of reviewers (unless in the latter case, there is convincing evidence otherwise).

Should an appeal garner an editor's consent for a resubmission, then she will provide guidelines in terms of what is expected and how the manuscript will be handled. Depending on journal policy, the editor will solicit reports from either previous and/or new reviewers. Similar to major revisions and soft rejections, doing your utmost to satisfy critical shortcomings does not mean that your resubmitted paper will be accepted. Editors will pay particular attention to whether the reviewers and editor are swayed by your revisions.

Appeal letters are usually sent in the days following a publication decision. An appeal sent a few weeks or more after a publication decision will probably not be considered. Journals are generally reticent to entertain appeals unless the underlying reasons are rock solid. Therefore, I recommend that appeals are discussed with co-authors before writing the letter, and then to revisit the letter and reasons for the appeal after a couple of days, before finally deciding whether to proceed.

20
Data Archiving and Sharing

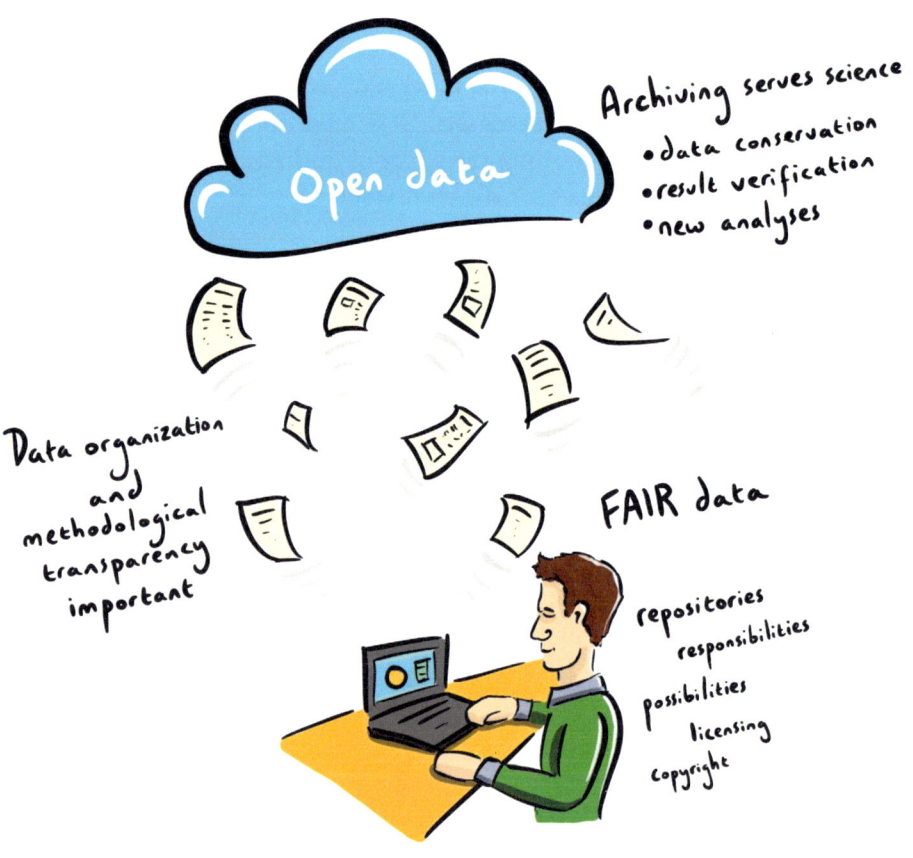

Many journals require data archiving as a condition for publication. Archiving permits verification of results and future use of data. However, data archiving presents some challenges, the most important being the analysis of data in new ways either not approved by—or potentially scooping—the original authors. To whom does data belong and what is fair practice for future use? This chapter will discuss these issues and ways forward.

One of the realities of science is just how challenging conducting and publishing a research project can be. Hundreds if not thousands of person-hours go into a study and generate sometimes massive amounts of data that are then analyzed and presented as findings in the final article.

The main issues of data conservation and future use were and still are:

- How to organize a data book and data files.
- How to conserve the books or files.
- How to ensure data survival after the end of professional activity.
- Under what circumstances can other parties access the data.

It is standard practice that a scientist keeps a data book. Data books and data files are central to science because they contain the original raw data and the methods used to collect them. This information serves both to accurately write the Materials and Methods and conduct the analyses that will form the Results. But the data also have a longer-term value. You may use them in unanticipated future analyses, and could be contacted by other researchers either to verify your analyses or conduct new analyses.

Such data—unless part of an ongoing study—will probably never see the light of day again once the PI is gone. A typical PI will store data books or files in a safe place until that day when, in leaving science, she passes them on to her successor, archives the data at her home institute or, alas, the data are not passed on and are irrevocably lost. Short of having the means to track data to their source, the first two of these alternatives are useless. Just like the third case, the data are lost in the long term.

When we realize that many thousands of hours went into organizing, collecting and collating a *single* career's-worth of data, the loss summed over all the world's scientists is unimaginably huge. Data archiving and data sharing address this fundamental issue.

Archiving and Sharing

Prior to the advent of repositories, data were self-archived. This usually meant no added effort on the part of the scientist. As mentioned above, data were already in books and electronic files. The scientist was morally prepared for requests from other scientists to verify published results. This could be anywhere from a data table to a fitted curve in a figure, to a statistical result, and even the verification of the coding of a computational model and/or the results it produced. Such queries were uncommon but always possible. Self-archiving was—and is still—a central tenet in good scientific practice. It was easy.

> *Whose data?* Numerous ambiguities relating to data permanence and ownership already existed in the pre-electronic era, owing to the lack of rules and guidelines. Like so many other aspects of writing and publishing science, archiving was learned either through supervisors, senior colleagues or on one's own. Data not only "belong" to the collector(s), but also in some respects to the scientific community (see below). Sometimes rights of use and ownership can be complex. For instance, in the US, the institution who commissions research is legally entitled to the data. Institutes and agencies may also restrict data use when potential revenue is involved (e.g. patents), and when data publication could have prejudicial ramifications (e.g. toxic effects of a supposed safe chemical).

But with the advent of the Internet and electronic publishing, and increasing demands for data transparency, availability and permanence, numerous initiatives were launched in the 2000s to foster the transfer of data to dedicated repositories; that is, archiving or "open data." This not only meant that data would not be lost for eternity, but that they could be rapidly accessed after publication.

Initially, journals adopting an archiving policy made it voluntary. Authors were encouraged to deposit data files at an independent, dedicated repository. Obtaining journal and author compliance has been a gradual process.[1] Today, many journals have made archiving mandatory and a number of article and research data repositories are now available.[2] Notably, in Green OA, authors used to exclusively deposit their accepted manuscripts in a repository, but this is now increasingly ensured by librarians and publishers.

Archiving introduces both responsibilities and possibilities.[3]

> *Where? What data? What format?* Archiving is of greatest service when formats are standardized and entries centralized. Both present challenges: establishing nomenclature and classifications for the former and recognizing that higher standards require dedicated—sometimes specialized—repositories for the latter.

> The minimum requirement for archiving is that any interested party can verify any result in the published paper. This introduces the issue of user-friendliness in the dataset and the responsibility of authors to provide enough detail so that any complex methodology for analysis in the original paper can be replicated. But it also exposes the related question of whether data collected during the execution of the study, but not used, should be made available to potential future users (see below).

> *Misuse and scoop not so important.* According to a recent survey by Stuart et al.,[4] the greatest barriers to data sharing are how data are organized, uncertainty in copyright and licensing, not knowing which repository to use and the lack of time to deposit data. Fears of misuse and getting scooped were of minor concern among respondents.

Reanalyzing data. Few things in a scientist's career could be worse than being notified that her results are in error. But errors do happen and, if significant, should be reported. Although journals may suggest how this is best done, there are no hard and fast rules. The prudent and diplomatic scientist will first approach the authors of the paper in question to report the anomaly. This not only prepares the authors for the eventuality of a correction (or if very serious, a retraction), but also gives them the opportunity to reply, eventually countering the inquisitor's contention. Should the contention persist, then it is ultimately brought to the attention of the journal where the original paper was published.

Future use. This is without a doubt the most controversial aspect of archiving, but also where the greatest potential benefit lies. Back in the day, scientists had control over the use of their own data, be it responding to a contention (which norms held them to take very seriously) or to a proposal that the data be used for other analyses. Should the data be released after the scientist ceased to be in activity, this was usually no issue for its use. With repositories, things are different.

If I have spent years collecting now archived data, is there a risk that others will propose doing analyses that I may or may not have intended to do at some future time? If I choose to release the data, do I have any say over their use? What prevents a scientist from scooping me with my own data?

Data sharing. The open data[5] movement seeks to conserve data and make them accessible to everyone. Open data has both advantages and concerns. *Advantages*: accelerates research; enables improved analysis of datasets; tests reproducibility; improves attribution of credit; conserves scientific information; and practical use in the applied sciences. *Concerns*: undocumented details of data collection; data misuse; original authors getting scooped. FAIR data[6]—findable, accessible, interoperable and reusable—are more restrictive than open data. Numerous platforms facilitate sharing.[7] Crossref "makes research outputs easy to find, cite, link, and assess." It facilitates links to the metadata of over 100 million records, including citations, books, journals, working papers, datasets and technical reports. Crossref connects articles via metadata but doesn't link to datasets themselves. Datacite is a registry organization that assigns a persistent identifier (DOI) to datasets, which Crossref then uses to associate with source articles. The Scholix initiative brings together more than 20 different organizations to promote the sharing of information and links between publications and datasets. Unfortunately, most datasets are not connected to their source articles with a persistent identifier, such as a DOI.

The main issue—which is being addressed by Open Science—is that data sharing is not always being rewarded. If datasets were given the same value as papers by funders and institutions, then concerns about sharing data would diminish.

Who has a say? Authors, journals, publishers and (increasingly) funders are involved in determining archiving and copyright options. We typically see one of two cases. In the first, copyright is completely waived and the data are in the public domain (Creative Commons Zero—CC0[8]). Users are free to do what they want with the data and need not cite the source publication (but users who uphold ethical standards typically will). A second case is CC-BY, which is similar to CC0, but users must credit the source. CC0 is usually preferred over CC-BY in large-scale metanalyses, since for the latter, every originating source must be credited. Best practice says that—for either option—authors should cite both the data and the article describing it (i.e. two citations), but—at least currently—all contributors to large datasets with 1000s of accession numbers cannot be credited.

The only protection authors of original work have is that data use can be embargoed by the source journal (typically 1 year).[9] This gives authors priority in reuse. Longer embargoes are possible but require special circumstances. In practice, scientists who are interested in verification of published results or reuse for their own analyses follow norms in best practices. This usually involves either proposing to collaborate or just citing the source publication.

PART V
CHALLENGES

Successfully writing and publishing takes concerted work and perseverance, and despite best intentions, things don't always go according to plan. You don't have the funds to cover article processing charges and so cannot submit your work to your first choice of journal. In submitting to another journal, you are disappointed by the decision, the time it took and the quality of the reviews. More generally, despite publishing regularly in reputable disciplinary journals, your work is rarely cited and your funding record is abysmal.

These are just some of the symptoms of greater challenges faced in the publishing world. As a citizen of science, it is important that you are aware of the issues, their causes and possible improvements. Many concerns are discussed in various sections of this book, but four in particular can be traced to what may be the biggest challenge facing science: the evaluation culture—the importance of how much and where we publish.

Chapter 21 considers an emerging threat to the scientific craft—the growing difficulty for journal editors to obtain appropriate, thorough, unbiased reviews. The problem is complex, involving the altruistic nature of review, anonymity in reviewing and the fact that scientists are oversolicited and ever-less willing to contribute. Chapter 22, looks at the cost of publishing, and specifically the market for scientific articles. Like any unregulated market, inequalities emerge and this is creating problems for article access and authors to cover article publication charges. Chapter 23 examines the raw material of evaluation—citation metrics. The formulae and the data upon which they are based have numerous shortcomings—we need to understand both what the numbers represent and how to use them responsibly. Finally, Chapter 24 focuses on "the science that never comes to be"—under-cited articles. Scientific papers live in a world of inequality, where journal names, gender, author identities and country of origin influence the impact of the science.

21
Peer Review

The main objectives of peer review are to improve manuscript quality and advise the editor regarding acceptance. But peer review's two defining characteristics—anonymity and altruism—are ill-suited to reliably fulfill its aims. Peer review has also struggled to keep pace with increases in manuscript submissions. This has led to a "tragedy of the reviewer commons," whereby over-solicited reviewers put less effort into reviews or stop reviewing altogether. This chapter presents problems surrounding peer review, and approaches to addressing the tragedy.

An Editor's Guide to Writing and Publishing Science. Michael Hochberg, Oxford University Press (2019).
© Michael Hochberg 2019. DOI: 10.1093/oso/9780198804789.001.0001

My first experience with peer review was in 1983 at UC Berkeley. I conducted a project investigating whether females of the California oak moth, *Phryganidia californica*, produced a sex pheromone to attract males of the species. Sure enough, this is what we found. Together with my supervisor, Dr Jan Volney, we wrote the study up as a research note for *The Journal of the Lepidopterists' Society*.[1] Before submitting the manuscript however, Jan suggested that we get "friendly" reviews from colleagues in the Department of Entomology. Three willing people reviewed what was a two-page, very straightforward manuscript. The ensemble of their comments was extensive, many dealing with the writing(!). These reports were invaluable to improving and ultimately publishing the paper.

These were not "real" peer reviews—they were our friends and colleagues—yet the experience had all the hallmarks of how and why peer review works.

Our reviewers:

- *Had the time*. They were likely not as solicited for reviewing as they would be today.
- *Took the time necessary*. They viewed a thorough review as important.
- *Were competent to review the paper*. Everyone we approached had the needed expertise.
- *Were constructive*. They viewed peer review as an important institution. Their identities were obviously known to us, putting an extra layer on the importance of carefully written critique.

Why these reviews worked then is also indicative of why peer review sometimes does not work today: scientists are busy and may give reviewing low priority. Some scientists ignore or refuse invites, do not take the time necessary for the review when having agreed and may hide behind the veil of anonymity to either adulate or destroy a paper. These and other problems are by no means givens—many if not most peer reviews *do* adhere to the positive points above. Nevertheless, there is a burgeoning issue that peer reviews do not always serve their purpose, and sometimes sully it.

Alternatives to classic peer review. Some are turning to preprint servers and subject repositories (arXiv, BioRxiv, peerJ, Zenodo, RePEc, PubMed Central) and websites with post-publication peer review (e.g. F1000, eLetters). These alternatives are not without their own issues; for example, the posting of embarrassing errors that are subsequently corrected (or not), and lack of clarity regarding whether the preprint is the final (citable) product. Another initiative is collaborative peer review (*eLife*, *Frontiers* journals, *EMBO*, *Science*), in which reviewers interact to produce a single report.[2] This maintains anonymity and promotes constructive, consistent and quality assessments.

Anonymity and Altruism

Although the ideal of peer review is to benefit science, and authors express support for peer review,[3] it has numerous shortcomings including: delaying publication, over-influence on editorial decisions, biased reports and the risk that manuscript information will be used somehow by reviewers in their own work.

Many of the concerns surrounding peer review stem from its two defining characteristics.

Anonymity. Not being known to authors means providing comments without fearing judgment or reprisal. Anonymity is intended to promote more honest reviews.

Altruism. Reviewing contributes to the scientific commons with the expectation that others do the same. But conducting a thorough, thoughtful review takes time that could be spent on one's own research.

Although many use anonymity as it was intended, others do not, or even abuse it. Similarly, altruism is predicated on one's willingness to contribute to the science collective. No "higher authority" checks to see if scientists contribute their fair share, meaning that invited reviewers are free to decline or, should they agree, conduct their (anonymous) assessment pretty much as they please. Fortunately, concerning the latter, the overwhelming majority respect the institution of peer review and conduct conscientious reports.

When problems do arise, they can take several forms:

- *Lackluster reports*. The reviewer conducts a summary or careless report.
- *Biased or "rogue" reports*. The reviewer biases her report either to support or lambaste a study due to, for example, author identities or schools of thought.
- *Delayed or cancelled reports*. The reviewer gives low priority to conducting, intentionally delays or never submits the review.

Editors can do little to prevent these issues beyond inviting people they view as conscientious in the first place and, if necessary, arbitrating problematic reviews. Journals are however increasingly sophisticated in their data analytics and keep track of various measures of reviewer performance. Reviewers who systematically decline to review, submit late reviews, or produce summary or biased reports are less likely to be invited for future assignments or will be placed on a no-invite list. As much as some editors may relish the thought, journals do not reprimand uncooperative reviewers.

Issues surrounding reviewer responsibility are covered in more detail in Chapter 28. The remainder of this chapter discusses how the over-solicitation of scientists is eroding altruistic peer review.

More Manuscripts—More Experts

According to the Web of Science Core Collection, 2,896,773 articles were published in 2016, and this is about double the number in 2003.[4] Publons[5] estimates that the annual growth in published articles from 2013 to 2016 was 2.6 percent, whereas the number of manuscript submissions grows by 6.1 percent annually. The number of individual authors is also increasing at comparable rates. According to one survey, their numbers almost doubled between 2003 and 2013.[6]

Because these articles are peer reviewed, it would seem therefore that the pool of reviewers is keeping up with increases in manuscript submissions. But according to the Publons survey the average number of invitations required to get one review increased from 1.94 to 2.38 over the period 2013–2017. Similar effects were found for a set of ecology journals.[7]

Why then are journals having greater difficulty in getting preferred reviewers to agree?

In an ideal world, each review received would be reciprocated to the scientific community at some time in the future by one or more of the co-authors. So, for example, if there were a total of 10 reviews received in four sequential submissions (i.e. at least three rejections), and eight authors on the paper, then (if the work were divided as equally as possible) six of the eight would review one future manuscript, and two would each review two future manuscripts.

But things are not so simple. There is no straightforward way to enable or enforce fair reciprocation. For example, if I am co-author on a publication, and "owe" one review, then how much time do I have until I conduct it? Is there any review quality control? What happens if I don't provide the required review?

Reasons for not pulling one's weight are many and varied. They include: not being invited to review papers in the first place, being invited to review papers outside of one's expertise, and being invited, but having conflicts of interest. And some people are just busy. Busy can mean personal or professional imperatives or prioritizing one task over another. Reviewing takes time and effort and is not directly recompensed. It therefore competes with other projects promising some kind of return (e.g. a published article). "Busy" can also be an excuse to think that one's review is not indispensable—"there are enough competent reviewers out there to do the job." Declining to review need not be justified, and is easy: just click on the appropriate link or ignore the soliciting email.

Some scientists are more likely to be contacted to review papers than others. The number of authorships, citations, appearances at conferences and the regularity of reviewing for journals, all contribute to being highly solicited to review. Experience-bias is therefore built into the system: students, postdocs and young faculty are less likely to be identified as potential reviewers, despite their reviews often being as or more thorough and useful than those from more senior scientists.

What overall effort goes into peer review? Consider the following simple but plausible calculation. Assuming on average a very low one review conducted per submission, this means that anywhere from one (accepted in first journal approached) to several (assuming two or three journals eventually approached[8]) reviews contributed to marshaling these manuscripts to publication. If the number is a high three reviews per submission, then this number can be anywhere from three to more than 10! The actual number of reviewers will likely be somewhat less than the number of reviews, since some experts are contacted more than once, following rejection by one journal and submission to another. Nevertheless, the total effort put into reviewing published (and sometimes never-published) papers is considerable and, arguably beyond several reviews conducted, becomes redundant and a waste of reviewers' time.

The Tragedy

Reviewing manuscripts is an altruistic act. It takes time and effort and means doing less of one's own work. It can also mean—in making extensive suggestions and corrections—that reviewers contribute as much (or more) to the final paper as some of the co-authors, but the former is, at most, only acknowledged.[9] Peer review works when scientists view this sacrifice as both a responsibility and an opportunity to contribute to the betterment of science. Some are also motivated by the expectation that other scientists feel and act the same. They indirectly reciprocate. They cooperate.

But for peer review to work, enough people need to contribute, and specifically the *right people* are needed. The right people are those scientists who, according to the editors, are competent to review specific manuscripts, are thorough and are reliable. The system suffers and eventually breaks down when the right people are over-solicited and decide to review less than their share, or nothing at all. When the right people contribute less to peer review, less appropriate reviewers take their place.

The "tragedy of the reviewer commons"[10] occurs when different journals *tend to seek the same individual reviewers*.[11] Each journal invites specific people for their own purposes, as if no other journal were inviting the same people at other times, for other manuscripts. This happens for the simple reason that editors never communicate their reviewer lists. Imagine the following plausible example. If I am the editor of journal X, then I may invite reviewer Y (who always accepts reviewing assignments) 10 times a year without knowing that Z other journals are *also* inviting Y on average (let's say) 10 times a year. The total number of invites this person receives in a year is about $10 \times (Z + 1)$, and as Z becomes large, one can easily understand that this favored reviewer simply cannot make the time to review all of these manuscripts! Worse, this person could become increasingly annoyed, and eventually accept only a few assignments, or may snap, and stop reviewing altogether for certain journals.

The tragedy affects journals as courts of quality science. Not only will substandard reviews result in less informed publication decisions, but when articles are finally published, they too will be at a lower scientific standard than had they been revised based on high standard reviews. Overall, the tragedy negatively impacts science.

> *Evidence for the tragedy?* Fox et al.[12] examined trends in the levels of invited reviewers agreeing and submitting their review for a sample of six journals in the fields of ecology and evolution. They found downward trends with time for some but not all the journals, upward trends in the total numbers of review invitations sent, no trends in the average number of invitations sent to each unique reviewer and that individuals invited repeatedly were less likely to accept reviewing. The authors interpret their results to indicate that reviewer fatigue only partially explains likelihood to review.

The tragedy affects science and the scientific community in a number of ways:

- More desk rejections so that journals can cope with fewer consenting reviewers.
- Emergence of journals requiring no reviews or lower scientific standards.
- Bias stemming from a smaller (willing) reviewer pool.
- Fewer, less useful reviews received.

The last of these points is particularly troublesome. Being asked to review a manuscript is recognition of one's expertise, and it used to be—and for many still is—a great honor. Nevertheless, being asked to review by a major journal such as *Science* or *Nature* carries greater kudos than receiving an invite from a less prestigious (but nevertheless reputable) journal. The obvious danger is that preferred reviewers are more likely both to accept invites by, and conduct high-level assessments for, prestigious journals compared with their lesser known counterparts. The latter will have greater difficulty in achieving quality publication decisions and improving manuscripts through peer review.

Slowing the Tragedy

The tragedy is a graded phenomenon—not all or nothing. Numerous checks slow the tragedy, and many journals apply these without even realizing the underlying mechanism that created the problem in the first place, or the greater positive implications for science.

Fewer go out for review

Until the 1990s, submissions that fell within the scope of a journal were routinely sent for peer review. Reviewers were plentiful and willing. Editors made it a point to give manuscripts their day in court. This situation changed with the growth of the Internet and more specifically the increasing importance accorded to citation metrics. The 2000s saw notable increases in submissions to the top-ranked journals and the first signs of reviewer fatigue and disengagement.

Journals increasingly found themselves in situations where they were reviewing manuscripts that had little or no chance of acceptance. They had to parse this with difficulties in obtaining reviews for those (viable) manuscripts where expert feedback was the most needed. Many journals responded to these challenges by increasing their desk rejection rates. Figure 21.1 illustrates the challenges of deciding between desk rejection and external review.

As unjust as they may seem, desk rejections have positive effects. They save time and effort for both editors and reviewers, and although disappointing for authors, enable a quick turn-around for submission to a more suitable journal. Desk rejections also help conserve a journal's reviewer base and result in lower costs since less editorial office time is spent marshaling peer reviews.

Revisiting the more extensive discussion in Chapters 12–13, in deciding on desk rejection, the chief editor—and often a member of the board—believes that the manuscript has little chance of surviving peer review. In making such a decision, editors have to rely on their experience in observing the range of manuscripts that are sent out for review at the journal and fail, and those that succeed. Desk rejections therefore necessarily have a subjective component and, in particular, solicit the ability of the editor to forecast whether a probable desk rejected manuscript could become an acceptable paper, were it to be revised. Of course, without expert reviews the editor is limited in making such forecasts, and it is for this reason that editors tend to err on the side of caution, and send manuscripts out for review where there is a doubt.

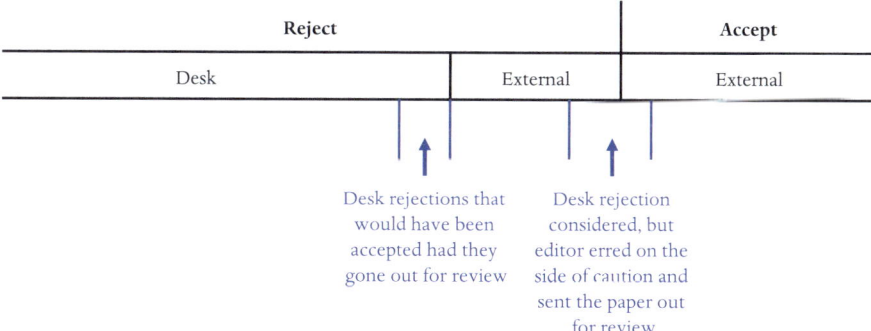

Figure 21.1 Editorial challenges in desk rejecting manuscripts. Had the editor correctly identified their potential, a small proportion of what were to be desk rejected manuscripts would have survived external review and been accepted for publication. Inversely, some manuscripts sent for peer review which are rejected, in hindsight should have been desk rejected. Proportions are hypothetical.

Nowadays, desk rejection is practiced by many journals. Although desk rejections undoubtedly reduce the tragedy, they are not a panacea, if for no other reason than because rejected authors can simply continue to submit to journals where they have little chance of acceptance, some of which will nevertheless peer review (and reject) their manuscript.

Submission fees

Although at first sight a reasonable policy to deter speculative submissions, submission fees are very rarely charged for biological science journals. The two main impediments are that they require administration and, more importantly, even if inexpensive, could become a significant factor in choosing a journal (see Chapter 16).

Invite younger reviewers

With age comes knowledge, and editors know this all too well. Given the choice between inviting a PhD student and a faculty member, most editors would not think twice and go with experience. But while expertise is invaluable, there are some notable downsides to systematically inviting more experienced reviewers. Some common problems include biased reports, summary reports and reports being late or never submitted. Over-solicitation of mid-and late-career researchers may explain some of these shortcomings, and a promising solution is to shift some of the reviewing responsibility to younger scientists.[13] This has the added benefits of training young researchers to review, accessing expertise in a hands-on career research phase, and getting greater punctuality and thoroughness than might be received from more senior (busier) researchers.

It should, however, come as no surprise that some young scientists are not equipped to conduct a peer review. Graduate school curricula very rarely include courses on writing scientific articles let alone reviewing. Journals, more experienced scientists and group PIs need to encourage, teach and train. This can be accomplished in two complementary ways.

First, journals should encourage the first steps in peer review. This is not to say that it's the journal's responsibility to take a student by the hand from start to finish, but rather that they identify prepared students and postdocs, and/or invite supervisors, requesting that they help their students take the first steps or gain additional experience.

Second, experienced researchers need to train students in peer review. Peer review is often overlooked as a part of a student's education, and rather is viewed as something learned "on the job." Learning on the job has its merits, but also its problems, such as untrained reviewers focusing on minor details and either wording their review too critically or congratulatorily. Rather, young scientists need to learn how to develop a critical eye while being constructive and courteous. In reviewing manuscripts, young scientists also become more aware of the strengths and shortcomings in their own writing. Experienced researchers can accompany students and postdocs by co-reviewing manuscripts or giving lectures or courses on writing and publishing scientific papers. Conducting peer reviews is treated in detail in Chapter 28.

Diversify the reviewer pool

Previously underrepresented scientific communities are increasingly publishing high-impact–high-quality science in reputable international journals. Yet, scientists from these communities are not contributing to peer review in the same proportion as their publications. The principal reason is *not* that they do not want to review, but rather that some editors are reluctant to approach scientists from unfamiliar communities or who are simply not registered in their databases. For example, according to Publons[14] scholars from China submitted over 17 million papers between 2013 and 2017, and were responsible for 13.8 percent of research output, but only 8.8 percent of review output. Signs are, however, that emerging regions—and China in particular—are increasing their contributions to peer review, with growth of 224 percent in 2017.

Ask dedicated editors to conduct reviews

Journals range widely in the responsibilities of editorial board members. Many journals keep editors out of the manuscript review process, preferring that they focus on arbitration tasks such as initial screening and recommendations on reviewed manuscripts. Contributions of editors to reviewing either punctually or regularly, will alleviate the tragedy.

Clearly, those rare journals that review exclusively "in house" are unaffected by the tragedy. Moreover, they reduce the tragedy by "shielding" considerably greater numbers of scientists from peer review solicitations. Editors-as-reviewers, however, potentially bring other issues into peer review, such as not being sufficiently qualified to assess certain manuscripts or in introducing bias.

Recompense, acknowledge, publish

Compensating reviewers is arguably the most effective way of curtailing the tragedy. It provides a just reward directly for each contributed review, rather than relying exclusively on the mechanism of altruistic, indirect reciprocity. Providing rewards should, in particular, promote quality reviews, since the service is under a costlier contract than just an automated "thank you," and the quality of the service rendered can be checked.

There are numerous non-mutually exclusive ways to acknowledge, reward or compensate reviewers, including:

- publish names at end of journal;
- PubCreds;
- lowered costs for subsequent publication or online access;
- recognition at home institute;
- open peer review;
- monetary compensation;
- professional reviewers.

None of these is routinely used and, although the last two have been widely discussed, I know of no journal that monetarily compensates reviewers. Paying reviewers is problematic because it requires administration and accounting. PubCreds, where reviewers earn tokens that are otherwise required for manuscript submission,[15] is really only viable if many scientists and journals participate. Publons allows reviewers to "claim" their reviews (a citable DOI), and the service ensures the right people get the appropriate credit.[16] Finally, the increased transparency of open peer review at OA platforms and journals is gaining considerable support.[17–19]

Cascade journals and revising rejected manuscripts

Journals that send what should be desk rejected manuscripts out for external review increase the tragedy. Over-reviewing is difficult to estimate, if for no other reason than because the justification for sending a manuscript out for review is rarely noted. It is the remit of the journal, and not the community, to consistently administer a peer review policy. Desk rejecting submissions fairly, requires editorial time and expertise, and is never applied to all eventually rejected manuscripts, since editors cannot perfectly predict the outcome of peer review (Figure 21.1). A manuscript with potential at one journal that is rejected following peer review will possibly experience the same fate at one or more subsequent submissions. Thus, journal independence and erring on the side of caution mean that some manuscripts are reviewed multiple times before eventually being accepted.

One partial solution to the burden of re-reviewing a previously rejected manuscript is to transfer reviews from journal to journal. Typically, this is done within a journal family under the same publisher, both to facilitate assessment leading to publication, and to promote the flourishing of new or lower standing members of the family (so-called "cascade journals"). The authors will be expected to reply to reviewer comments from the original submission. Some non-family journals that consider reports from a previous submission will nevertheless solicit one or more new reports.

> *Cascade journals.* Many journals now live in a "family" of titles under a single brand name. This both keeps excellent papers within the family and helps promote newer family members. Inheriting rejected manuscripts from more famous siblings gives an automatic boost to young journals, since more submissions permit more selection on quality and interest, higher reputation and impact, more readers/subscribers and higher revenue. Examples of cascade journals include *Nature*, *Nature Communications* and *Scientific Reports*, and more disciplinary families such as Wiley family journals and *Ecology and Evolution*.

There are reasons why some authors are not keen to transfer reports from a rejection decision to a new journal. These include (obviously) not prejudicing manuscript handling and avoiding the possible obligation to address what the authors feel are unreasonable or erroneous reviewer comments. Even if not informing a newly approached journal of a previous submission (which is the author's perfect right), taking previous reviews into serious consideration in revising a manuscript will improve it and increase chances of subsequent publication success.

22
The Cost of Publishing

The cost of publishing is hotly debated. Until the 1990s, publication was free of charge, paid per page or per article, or covered partially or completely if subscribing to a journal or if a member of the academic society overseeing a journal. A major shift occurred in the early 2000s when new "Open Access" publishers made articles freely available for all to read and reuse, with article processing charges being covered by the author. This chapter discusses the paradigm shift and how it has changed the landscape of who pays for scientific publication.

The good old days! When I was a student in the 1980s the choices of which journals to browse were simpler and the cost of subscriptions generally covered by academic departments. The library would have both discipline journals and "vitrine" journals such as *Science* and *Nature*. The most recent issues would be on display and older volumes could be accessed in the library stacks or through inter-library loans. The department's budget was nevertheless finite, meaning that if you wanted rapid, frequent access to journals not covered by the institute, then you would need personal subscriptions or access to a friendly colleague's collection. If you were interested in a specific article, you could send a postcard to the corresponding author and receive a nice, shiny reprint. Thus, if you were lucky enough to be based at a well-funded institute in the west, then—with some sleuthing—you had access to the vast majority of scientific publications, and typically paid annual fees of no more than a couple hundred dollars in personal subscriptions.

While someone or some institute had to pay for the right to read articles, the cost to publish was generally free—but with some exceptions. For example, some society journals would charge publication fees for non-society members. Many journals published a fixed maximum number of pages per year, meaning that authors had to write concisely. Papers either were not allowed to exceed page limits, or if they did extra fees were charged.

In the end, paying to read or paying to publish was almost never an issue for a lucky scientist.

Those days are gone. Many readers and institutes cannot cover article access costs at paywalled journals, and many authors do not have the funds to even publish in certain high-profile journals.

> *Journals live in a competitive world.* Over the first 200 years of scientific publishing from 1665 to 1850, journal numbers increased by two orders of magnitude from one (*The Philosophical Transactions*) to about 100. From 1850 to 2018, the numbers gained more than two additional orders of magnitude from 100 to over 10,000. The continued radical expansion in journal numbers owes to the expansion of scientific disciplines, more people doing science and greater productivity per researcher. More papers submitted to a fixed number of journals means higher rejection rates, and a market for new journals to absorb some of these manuscripts. Starting a new journal is very costly and, unless there is a "niche" for the journal project, it is unlikely to attract sufficient numbers of exciting manuscripts to be sustainable for the journal and profitable for the publisher. Most ideas for new journals never get off the ground.

The Shifts

In the 1980s and 1990s there was a gradual shift from single-department and single-person subscriptions to consortium subscriptions (so-called "Big deals"). In a consortium, multiple institutes associate and gain access to a bundle of journal titles. The grouped journals typically covered the majority of an institute's (now redundant) subscriptions. Access to published research increased immensely from the late 1990s into the 2000s. This was great for the individual scientist, since not only were many of her regular reads included, but so were a number of previously inaccessible, but potentially important journals. Despite an overall increase in journal access, many specialized journals were not included in these lists, meaning that individual and departmental subscriptions were still necessary. Bundling journals nevertheless proved to be cost-effective and very profitable, and this system still flourishes today.

Even with the efficiencies of consortia, the basic economic model turned out to be vulnerable. Personal subscriptions were paid by the reader. Departmental and consortia subscriptions came from one or more of a variety of sources including governmental agencies, university funds and private foundations. Subscription prices continued to increase despite the shift from print to electronic publishing, and big deals meant that libraries could not drop individual journals at the expense of access to new journals. Access to publications increasingly became an issue, both because of paywalls and that authors had little say, since they generally did not hold the copyrights to their own papers.

> *The serials crisis* is the increasing costs incurred by publishers not being offset by the ability of academic libraries and their funders to cover the subscription prices. The two most important factors contributing to higher publisher costs and subscriber prices are the increased research output resulting in the growth of article number per journal and journal numbers (in small, costly, niche areas) and price inflation. In parallel, academic library budgets and even well-funded institutes have not been able to keep up with annual price increases and subscriptions to new journals.[1] The "one article—one payment" philosophy of APC OA has been developed largely in response to the serials crisis. Nevertheless big deals continue: big subscription deals are now being replaced by big OA deals.[2]

In the early 2000s BioMed Central and the Public Library of Science, or PLoS, proposed an alternative to subscription access and article copyrights, the latter typically held by either the journal or its publisher. The focus was on costs and rights of being an author or a reader. Authors would pay for APCs, rather than departments paying for subscription charges. Articles published with what has come to be called Open Access (OA) are "owned" both in terms of copyright and article distribution by the creators; that is, the authors. This turns the economics of publishing on its head: those whose interest it is to diffuse their research pay for it; those who wish to read it, can do so for free. In the subscription model—sometimes called "toll access" or "reader pays"—the subscriber pays regardless of whether she reads.

One would think that this should simplify things. After all, a single party (authors) covers one-time charges and everyone has access. Rather than being simplified however, the economic landscape in publishing has become more complex. Several models (presented in Chapter 15) now co-exist in terms of author and reader rights and who covers charges.

Subscription only. This was the norm until the advent of OA. Individuals or institutes subscribe to the journal and *only they* have unlimited access to its contents. Non-subscribers can purchase individual articles. There is no cap on publisher revenue meaning the more subscriptions, the higher the total revenue.

Gold OA. These articles are exclusively OA (CC-BY) and are published on the journal website. Any APC is charged to the authors (only if their manuscript is accepted for publication), and usually paid by a personal research grant or academic department. When there is an APC, authors from designated low-income countries either pay reduced charges or publish for free. Some Gold OA journals have no APCs, though these tend to be regional journals, or specialized, low-impact venues. Published articles are immediately accessible via the journal's website and authors can post and distribute copies of the article as they please.

Membership OA. A recent initiative by *peerJ* is to give authors the option of either covering APCs (much like in Gold OA), or becoming a lifetime member. To use the latter option, all co-authors of the submitted paper must be lifetime members. Depending on the membership level, the author has the right to publish up to five peer reviewed articles per year. Should the membership option not be possible, the APCs are competitive compared with other Gold OA journals.

Green OA. These are subscription-only journals that allow the distribution of accepted versions of a manuscript, usually only after an embargo period—depending on the publisher—of anywhere from 6 to 24 months. Either journals or authors themselves are responsible for depositing their paper on a non-commercial server, such as PubMed Central. Copyright conditions vary from journal to journal.

Hybrid journals. Some subscription-based journals offer Gold OA to those authors willing to cover APCs.

The publishing ecosystem has therefore become complex and largely unregulated economically. There are a number of problems and questionable practices that remain to be resolved:

1. Although overall access to scholarly journals is at a historic high, many scholars based at large consortia-subscribed institutes cannot gain access to certain journals, particularly those outside of the institute's academic disciplines. Researchers in developing countries can gain free or low-cost access via Research4Life programs,[3] although this concerns only a subset of countries with very low GDP, and many institutes in high GDP countries don't have the support to gain access, and do not qualify for low-cost access.

2. There is little to compel a hybrid OA journal to adjust its pricing for either OA or subscriptions[4]—so-called "double dipping." Moreover, APCs in hybrid

journals are, on average, significantly *higher* than those in Gold OA journals.[5] The majority of consortia are aware of these issues[6] and increasingly use them in "offsetting" agreements.[7]

3. For those OA journals charging publication fees, authors who cannot afford to pay, cannot publish. There are three main issues surrounding author responsibility to cover APCs. First, authors in higher-income countries are liable for APC costs, irrespective of whether they have any funding, and any funding they should have is earmarked for that purpose or not. A journal has *no way* to verify an author's funding situation. Second, and similarly, funding begets research, and high-impact research begets more funding. This creates a barrier to underfunded groups that produce what *would be*—but does not actually become—high-impact research, because costs cannot be covered in the high-impact journals.[8] And third, by charging high APCs, journals can reduce submissions from less competitive research groups, both meaning fewer potentially low-quality manuscripts to handle, and selection for papers from the most exciting research groups (which have the funding to cover high APCs).

APCs and subscription costs that impede manuscript submissions and published article access, respectively, do a disservice to scientists and to science.

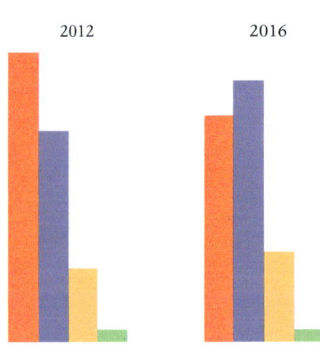

Figure 22.1 Growth of Open Access (OA) journals from 2012 to 2016. Bars show relative proportions of each of four models. Red = subscription only; blue = hybrid (subscription and Gold OA in option); gold = Gold OA; and green = Green delayed OA (though licenses vary). (Based on data from Scopus.)

Plan-S is an initiative launched in 2018 by Robert Jan Smits and a group of 11 European science funders to eliminate public paywalls in science. The plan aims to[9]:

Ensure that by 2020[10] all articles will be immediately available as OA, with authors retaining the copyright.
Eliminate hybrid OA.[11]
Publication charges will be paid directly by funders and APCs will be capped.
Funders will promote OA publishing and platforms.[12]

What Goes Into Pricing?

Journals—and therefore the scientists who publish in them and read their papers—increasingly live in a world of economics.

The annual revenue generated by scientific journal publishing has been estimated at about $10 billion, with annual price increases of 6.7 percent from 1986 to 2011 compared with a yearly 2.9 percent increase in the US Consumer Prices Index.[13] Estimates of profits before tax are higher in commercial vs. university press or society journals,[14] with the former attaining double digits and as high as 40–50 percent according to some sources.[15,16]

APCs and subscription prices vary anywhere from hundreds to thousands of US$, and not surprisingly tend to be less expensive in academic society journals compared with non-society journals.[17] The actual costs that form the bases of these prices and for subscription journals are largely unknown,[18] but may stem from many activities, including editorial and production office staffing at the journal level, and technology spending and investment at the publisher level.

> *Costs and prices.* There is considerable variation from publisher to publisher in prices and from journal to journal in what the consumer pays for access.[19] Two of the main sources of variation are the cost of labor (ensuring peer review, and whether labor is outsourced) and salaries/honoraria paid to editors and editorial staff, but numerous other tangible and intangible costs enter into the final figures that can be anywhere from hundreds to thousands of US$, with levels being on average about 50 percent higher at subscription-based compared with OA journals.[20] APCs per article vary considerably from journal to journal, from about $10 to $5000, whereas revenues per article in subscription-based journals are on average about $5000.[21]

Academic society journals typically price based on expenditures and reinvestment, but can also make considerable profits (called "surpluses"). Surpluses are invested or used to support activities such as seminars, conferences, research grants, prizes and public education. Publishing company journals price according to these same two factors, but with a larger emphasis on what the consumer (subscriber or author) is willing to pay, which is influenced by reputation and impact factor[22] (see below). Companies invest profits in research and development and, if applicable, distribute some of the profits as dividends to shareholders. The distinction between what only 20 years ago were accurately called "not-for-profit" and "for-profit" journals is becoming blurred, since academic society journals are increasingly publishing with "for-profit" publishers and sharing the surplus.[23]

Pricing is also affected by the position of the journal in the market. Many small publishers—particularly not-for-profit—have difficulty growing in a market dominated by a handful of large, for-profit publishers. This is particularly significant in contrasting the prevailing seller's market of journal subscriptions with the buyer's market of newer OA journals, and both the reluctance of the former to flip to the latter, and the difficulty for the latter to penetrate the market.[24] A young journal without a major brand name will likely start either free or with a low APC (typically hundreds of US$), though those that are part of brand names can price at higher levels. Nevertheless, a young journal has "sunk costs," and some attempt to recover these through higher pricing. As journals age, they gain a reputation and demand increases (submitted manuscripts). Prestige means ensuring more time and effort goes into each submitted article, but larger operations can to some extent lower per article costs due to economies of scale. Both journal rank and internal costs could justify increases in subscription prices and APCs; rather, prices and price increases are built into APCs and consortium and library subscription deals and based, in part, on willingness to pay.[25] The fact that consortia do not have access to the prices paid by other bodies ("non-disclosure deals") means that price discrimination can be easily practiced.[26] Price sensitivity in the subscription market is low, because intermediaries such as consortia and libraries negotiate contracts rather than the consumer (scientists) themselves.[27]

Many of the top-ranked journals are either purely subscription-based or hybrid OA. These journals have a considerable subscriber base, making it economically infeasible to shift to complete Gold OA.[28] Prestigious subscription-based journals are also hesitant to shift to Gold OA since APCs could pose a barrier to some authors submitting high-impact papers.

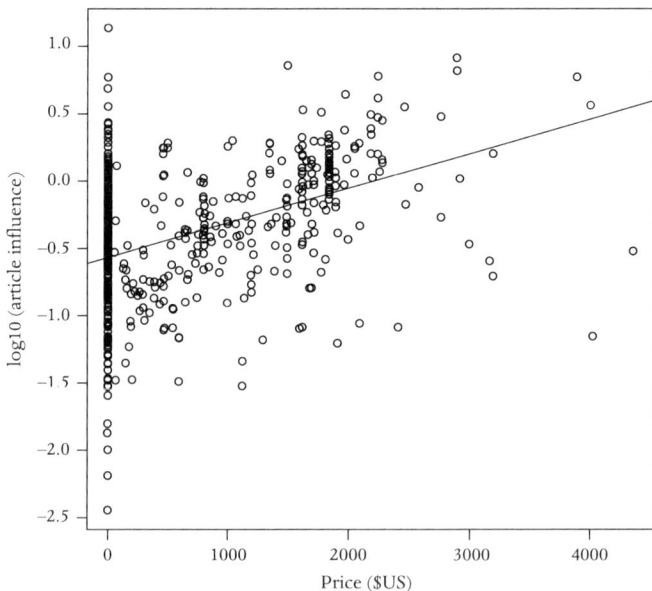

Figure 22.2 Correlation between article processing charge and Article Influence score for items covered by Journal Citation Reports (West et al. 2014).

Predatory journals prey on authors that are unable to publish in reputable titles. These journals exploit the market by removing barriers: their APCs are low and scientific standards necessary for publication are minimal or non-existent. Many of these journals are effectively "clearing houses" that receive manuscripts, convert them to a journal format and post them online. Given low running costs, these journals are able to maintain very low APCs and still make a profit.

An Uncertain Future

The publishing landscape is complex and not showing clear signs of resolving important issues.

Many of the downsides associated with unregulated markets are becoming apparent in scientific publishing. Primary among them is that an ever-smaller number of ever-larger companies dominate the scholarly publishing landscape,[29,30] and pricing reflects both the ever-greater dependence on these companies and what the client is willing to pay for reputation or prestige. A related issue is scientific inequality, with well-funded research groups having greater access to high-impact–high APC journals. Less-funded groups—regardless of the potential impact of their work[31]—may be unable to emerge from lesser-impact titles, and in the extreme, may be relegated to publishing in low-cost, specialized or even predatory journals. Plan-S and OA2020 aim to regulate APCs, but many challenges lie ahead.[32]

Central to these issues is the evaluation culture, and the necessity for (and often drive of) scientists to publish in high-rank journals. Initiatives like DORA and the Leiden Manifesto seek to reduce or eliminate the importance of *where* we publish and increase that of *what* we publish. Considerable work remains before our culture credits academic prestige to the scientific articles themselves. As long as indicators of prestige are associated with journals themselves, publishing companies will have a market and perpetuate the evaluation culture.

Perhaps the future lies in the innovation of academic society journals. Many society journals post accepted articles online, which speeds up dissemination. These journals could go a significant step further by establishing their own preprint services, together with post-publication review. Linking with providers such as Peer Community in . . . would not only maintain the positives of peer review and turn preprints into publications, but would significantly cut costs. Such a model, where society journals become service providers (OA, quality review and rapid publication) in association with not-for-profit publishers, would give scientists more say in the different phases of publishing.

23
Use of Citation Metrics

The idea that curiosity and discovery can be transcribed into numbers or rankings strikes at the very heart of science. The emerging norm of replacing scientific content with numbers is leading to an "impact factor syndrome," whereby metrics influence both where interesting and important research is published and how committees evaluate scientists and science. This chapter discusses different citation metrics and why the scientific community increasingly looks to them as measures of impact.

My first encounter with citations and impact factors was during a visit to the University of Kansas in 1994. The university had access to the Science Citation Index (SCI), something that my university in Paris did not. During the stay, I often used the SCI to conduct literature searches, since it was faster and more systematic than browsing abstracts in the library. But SCI had much more than just simple keyword searches: with just a few clicks it could produce an annual tally of the number of times any author's work had been cited. I was intrigued: I could see all papers that had cited my work, or anyone's for that matter. These data were complete and official. I recall that the total number of published papers citing all of my work was about 100, which I thought was pretty good . . . until I started summoning the numbers of colleagues at my own career stage. A bittersweet introduction to the world of impact!

Citation metrics gauge the importance of science at most any scale: a researcher's work, a specific article, a journal, a research institute or a country. Metrics are designed to use available data to quantify what is considered to best reflect some facet of impact. Because of the complex underpinnings of even the simplest indices, they are inevitably open to interpretation. Some—such as the Journal Impact Factor (JIF)—are the subject of considerable debate.

This chapter is about the uses and abuses of numbers.

What is Impact?

Impact is how science and scientists are changed by an article or specific findings in an article.[1] The accepted unit of scholarly impact is the citation, because this is the only concrete evidence of how a source article potentially influences readers. There are numerous issues with measuring impact using citations. First, there are different reasons for citing articles. A paper cited as part of a group of articles to give readers links to further information is much different from one cited as being the basis for a study itself. Second, authors may cite a given article one or more times, but this will only be counted as a single citation in most metrics.[2] Third, and importantly, there is no simple way to associate how reading or citing an article changes a scientist (her research) or science, respectively.

Journal Impact: The Journal Impact Factor

The first measures of journal impact emerged over a decade after the more modest proposal by Eugene Garfield in 1955 to group influential journals and the articles within into an "index." Garfield argued for the importance of the traceability of papers citing previous work. By examining a list of articles citing a target paper, one could learn what kinds of refinements or refutations had occurred in the intervening period. Garfield founded the Institute for Scientific Information in 1960, which produced the first science citation index in 1963, aimed to improve the scientific process and not to assess impact *per se*. Nevertheless, it provided the basis for the invention of the JIF and its first publication in 1969.[3] The index and its products became the premier reference in 1997 with the launch of the Web of Science (WoS).[4]

The JIF[5] for a given journal in calendar year Y is calculated as the number of citations received in calendar year Y for items covered by the source database, published in years $Y-1$ and $Y-2$, divided by the total number of citable items published in these same 2 years. This calculation necessarily has bounds. They are placed on the census period (calendar year Y) and the items being censused (years $Y-1$ and $Y-2$). Moreover, all items from these two previous calendar years have equal weight, despite up to 2 years separating the first from the last publications. In other words, the earliest publications from the period have had two additional years to be read and cited compared with the last ones. All primary research and review items are included as citable in the denominator, and these *plus* other more peripheral rubrics such as editorials, corrections and commentaries are counted among the citations in the numerator. Finally, all else being equal, a journal that tends to publish—*a priori* more cited— review papers will have a higher JIF compared with journals exclusively publishing— on average less cited—primary research.

The founders of the impact factor were conscious of some of its limitations. Their goal however was to produce a single number that reflected a journal's influence. Let's briefly explore the main elements in the calculation of the JIF, and in impact factors proposed by other authorities.

Journals covered

Determining which academic journals merit coverage is not an exact science. Depending on the authority, there are between 10,000 and 30,000 citable scientific sources. Most authorities use the same core set of criteria but apply them to different degrees when determining whether they include a given journal. More exclusive providers retain only journals of the highest standard. They apply more criteria and are more demanding for each. Among the well-known providers, the WoS is more demanding than Scopus by Elsevier. Metrics by Google Scholar (GS) has very few quality restrictions on what they cover (roughly three times more documents indexed than either WoS or Scopus). Moreover, because GS only indexes records on the Internet, it lacks the historical depth of WoS and, to a lesser extent, Scopus. Finally, WoS and Scopus are elaborate, subscription-only providers, whereas GS is less developed, but free.

More selective does not mean better. Whether an article is covered by a database depends on the journal in which it is published and not article quality or potential impact. Thus, great, potentially impactful papers may be omitted, and faulty, mundane papers included by simple virtue of their host journal.

> *Factors influencing journal inclusion* at the WoS include: international reputation of the editorial board, association with a recognized academic society, submissions subject to peer review and the regularity of publication. Selection criteria and procedures can be complex[6] and their application opaque.

Period of publication

The period of publication can have a considerable influence on the numerical value of the impact factor. Earlier publication means more time for articles to have entered the scientific community and be cited. Wider covered publication intervals include more papers and therefore yield estimates that are less influenced by small numbers of very highly cited papers. The JIF is based on a 2-year stretch of publications that have had 1–3 years to be seen and cited. As mentioned above, obviously, papers published at the end of the 2-year period have had little time to enter the scientific community, and as such probably only weakly influence the calculation. To reflect longer time stretches and a larger sample of publications, Journal Citation Reports also calculates a 5-year impact factor: articles published from $Y-5$ to $Y-1$, cited in Y and divided by the total number of items over the 5-year period. A concern with longer sample time spans is that the oldest items may no longer reflect the average importance of papers published in the journal.

Period of citation

Likewise, the *period of citation* influences the impact factor, but its bounds obviously depend on the period of publication. Thus, for the JIF to be released in 2020, the period of citation will be the 2019 calendar year and the publication period 2017–2018. If the citation period were to be extended back 1 year to also include 2018, and if a condition is that all published items have a chance at being cited during the full observation period, then the publication period would need to be rolled back at least 1 year (e.g. to 2016–2017). Similar to the period of publication, the JIF calculation becomes increasingly less relevant to the present as the period of citation is extended into the past.

Journal rank is the perceived prestige of a journal or the actual JIF ranking within a discipline. Rank notably eliminates absolute and relative magnitudes in average numbers of citations between disciplines and between journals within a discipline. Thus, for example, journal X1 may have a JIF = 10 and journal Y1 a JIF = 2, but both are the top ranked journals in their respective disciplines. Journal X2 may have a JIF = 2—8 points below X1, but nevertheless be ranked second in its discipline.

Individual Impact: Total Cites and *h*

All readers of this book will know that scientist too have impact. The two best-known metrics are total citations and the *h* index. Total citations is simply a tally of the number of times papers you have (co-)authored have been cited. Like the impact factor, total citations depends on the journals covered by the tallying authority. Your total citations are likely to be higher on GS compared with the WoS or Scopus.

The Hirsch or *h* index was devised to gauge both productivity and impact.[7] *h* is calculated as the maximum number of papers X that have been cited X or more times.

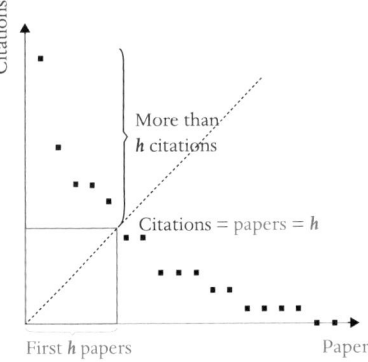

Figure 23.1 Calculation of *h* index.

Thus, I might be an author on 100 papers, 99 of them never having been cited and one paper cited 1000 times. It is also possible that I have only published one paper and it's been cited either once or 1000 times. In all three of these examples, my *h* is 1.

Like other impact factors, individual metrics have shortcomings. Perhaps the most important is that they are highly dependent on career stage. Careers usually take off in the years following the completion of the PhD, meaning that not only does a young faculty member usually have more published papers than a newly awarded PhD, but the former is publishing more articles each year, has had more time for her articles to influence the scientific community (and be cited) and likely has a more established scientific reputation. These elements indicate that it can be very misleading to use citation metrics to compare researchers at different stages of their careers—indeed even when comparing researchers at the same career stage. Imagine two people of the same age group, Betty and John. Betty has one first-author paper published 2 years ago that has been cited 100 times and John has ten papers published over the past 5 years, one of which has been cited three times, two twice, and the other seven, zero times. John's *h* is greater than Betty's, but some might argue that Betty is more likely to have a greater impact on science.

> *h is both paper and citation hungry.* There is a statistical relationship between total cites and *h*, such that *h* increases (approximately) with half the square root of total citations.[8] This means that *h* increases rapidly with citation number early in a career, but each additional citation's contribution becomes smaller as *h* grows. Because *h* is insensitive to a single or small number of very highly cited papers, three numbers could be necessary to represent a scientist's impact: one for highly cited work (e.g. the three most highly cited papers), one for the regularity and mass in citations over a career (*h*) and one that corrects the latter for career stage (*h*/number of articles published).

Some Other Measures

The many shortcomings of the most commonly used citation metrics have fueled a growth industry of alternative and complementary metrics.[9] Some of these are specific to journals, articles or individuals, whereas others (e.g. Citation half-life) can be applied to more than one category.

Clarivate also produces three other metrics:

The *5-year impact factor* is similar to the JIF, but calculated over the five most recent years of publication instead of 2 years.

The *Immediacy index* is a measure of how rapidly the average publication in a journal is cited. It is calculated as the number of citations of all articles in the journal in the year of publication, divided by the number of articles in that year.

The *Citation half-life* is the median age of all the citations received in a reference year. This is a measure of the impact longevity of a paper.

GS produces the i10 index, which measures the number of publications with ten or more citations. Similar in some ways to the h index, it quantifies the reliability of impact.

The Eigenfactor and Article Influence Score are impact factors that give more weight to authoritative sources in quantifying impact.

Influence and Beyond

As mentioned above, "impact" is the influence of science on science. The proxy measure for impact is the citation. Citation implies that a paper was read in the first place, and indeed, as discussed in more detail in the next chapter, many papers are read without being cited (and some papers cited without being read!). A widely read paper that is not highly cited could conceivably influence science more than one that is less widely read, but more cited.

Encountering but not necessarily citing scientific material has become known as *influence* and their measure as *altmetrics*. Examples include: visiting an article webpage, downloading pdfs and relaying information about an article on social media. Measuring *influence* is a burgeoning area. For example, the Google PageRank algorithm quantifies interest in published information based on the number of encounters and influence of the websites through which encounters occur. PageRank has been adapted to measure article impact and individual researcher impact.[10]

Spurred by the trove of information that could be used to gauge influence, several dedicated companies have emerged to create new influence indicators. For example, DigitalScience produces Altmetric, which incorporates "peer reviews on Faculty of 1000, citations on Wikipedia and in public policy documents, discussions on research blogs, mainstream media coverage, bookmarks on reference managers like Mendeley, and mentions on social networks such as Twitter."[11] Altmetric notably tracks the number of items published at each source.

The Allen Institute for Artificial Intelligence produces the Semantic Scholar, which ". . . helps researchers find better academic publications faster. Our engine analyzes publications and extracts important features using machine learning techniques. The resulting influential citations, images and key phrases allow our engine to 'cut through the clutter' and give you results that are more relevant and impactful to your work" (from FAQ on the Semantic Scholar website). Some of their projects include weighing apparent reasons for why papers are cited, and accounting for the number of times they are cited in a given document.

But despite considerable attention garnered by influence metrics, they suffer from some of the same shortcomings as impact. For example, it is impossible to know how influence actually affects science. The fact of encountering a website does not mean that the contents were carefully read (or read at all) and, even if read, what effect it may have had on a researcher's science. Equally important, levels of authority are not factored into measures of influence. People with absolutely no scientific credentials can post, access and relay scientific information. This means both that the influence score will include misinterpretations, and that the science most likely to have high influence scores will relate to the interests of non-experts.

The Impact Factor Syndrome

Despite certain citation metrics being available for almost 50 years, it is only over the past couple of decades that these numbers have had an appreciable influence on journal status, careers in science and on science itself.[12] Indeed nowadays, single metrics influence which journals attract which papers (the JIF), and which scientists get jobs, grants and the best students (the h index).

Why have these numbers become so important? Consider the following realistic scenario.

I am applying for research position with 20 other competitive candidates. The position is an open call in a department that spans all subdisciplines in biology. The committee is composed of 10 faculty members representative of the department's research. Each member is somewhat specialized and will be challenged to judge candidates outside of their own field. This specialization means that, when we look across the 20 applicants, we find that 16 of them would have at least one person who could judge their dossier, but four have no-one in their area. This means that—at least for the 16—all members have to "trust" the views and sincerity of those few who evaluate each dossier.

How can the committee competently judge something so important when there are no, or only one or a few specialists to seriously evaluate each candidate? How do they weigh publication number, authorship, the journal of publication, invitations to give talks, reference letters etc. in judging a candidate? How does a committee compare a paleontologist with a molecular biologist?

There are no simple, satisfactory answers to these questions. Each criterion used has a degree of arbitrariness and subjectivity, and indeed *subjectivity* is a key reason for why committees use citation metrics.

Despite their numerous shortcomings (see below), citation metrics are "objective" in the sense that they are based on formulae and data.

Metrics were a godsend for committees. They meant less time and effort in evaluating documents that were outside of their area, less discussion, and actual estimates of importance and impact. In a sense, the metrics shifted some of the decision to the candidates themselves (their science) and the scientific community (citers). Not only could numbers be compared between candidates with similar scientific backgrounds, but they could be used to compare a paleontologist with a molecular biologist!

Given their accessibility, apparent simplicity and meaningfulness—not surprisingly—citation metrics are now a mainstay of research assessment and management.[13] As a consequence, scientists have altered both the way they do science and where they publish their science. The power of committees—the "evaluation culture"—and the willingness of some scientists to acquiesce to or even embody the new normal of prioritizing impact has led to a lock-in: an "impact factor syndrome."[14]

> *Big data.* Academic institutions, funders and governments are interested in evaluating the quality, influence and impact of researchers, research programs and institutes. A number of analytics companies and initiatives now offer services based on bibliographic data and information about researcher and research group inputs, activities and outputs. Big data offers numerous possibilities but presents many challenges including quality, interpretation, reliance and use.[15]

Issues with Citation Metrics

Citation metrics are devised to estimate different facets of performance. They can convey useful information when used carefully, but unfortunately many users are not aware of their limitations. These include:

The metric. Each metric aims to quantify a concept of impact or influence. Some such as Elsevier's CiteScore favor their own journals over those by Springer-Nature.[16] Most indexes have a temporal component, either in the sample of papers or in the period over which they are cited. Each index returns a single number that may (grossly) misrepresent the underlying values should the variance be high or the distributions be skewed.[17]

The authority. A calculation reflects the data upon which it is based. Data used by Clarivate Analytics are highly curated, whereas GS employs a wide swath of sources. This means for example that researchers tending to publish in journals not covered by the WoS would have higher GS *h* indexes than WoS *h* indexes.

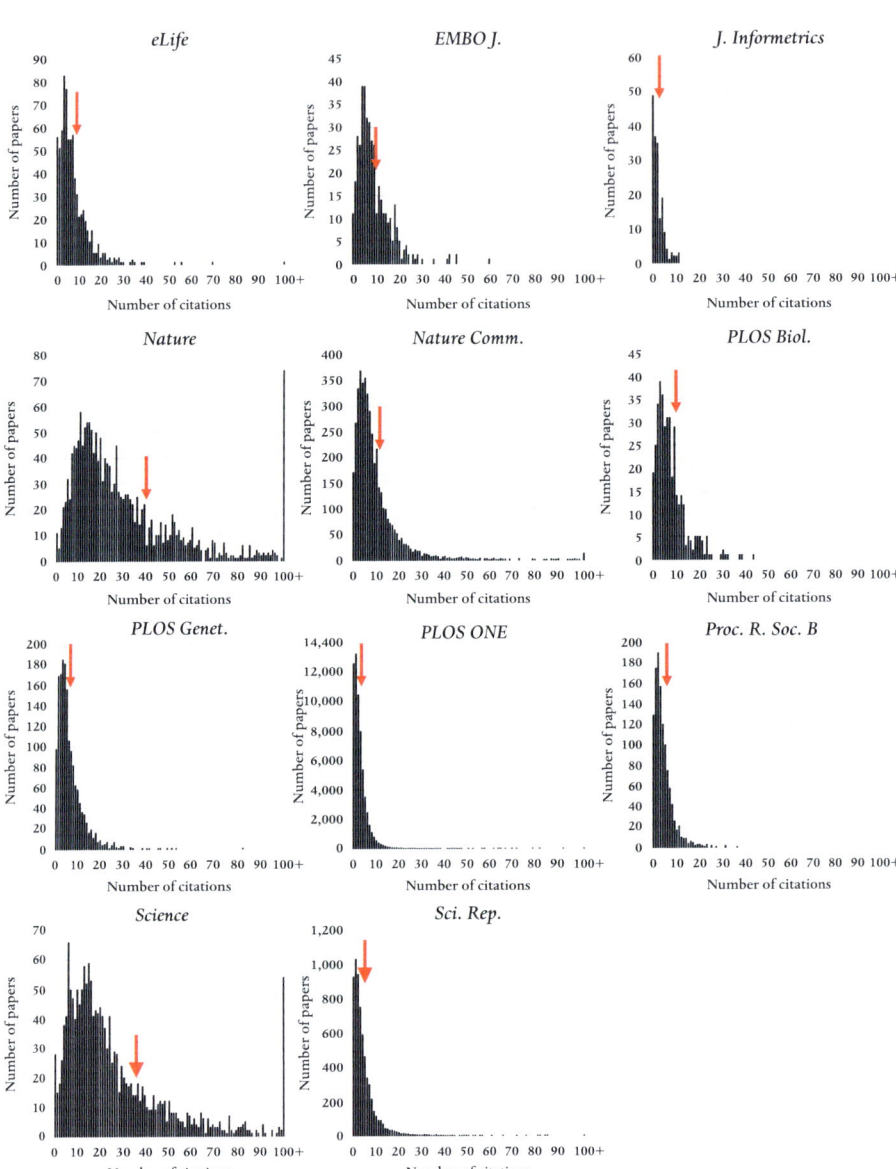

Figure 23.2 Citation distributions of major disciplinary and interdisciplinary science journals. Red arrows are added to original display (Larivière et al. 2016) showing approximate JIFs.

Stability. An index for a journal publishing 1000 papers annually will be more stable year to year than a journal publishing only 10 annually.

Negotiations. There is some evidence of impact factors being negotiated between publishers and the then Thompson Reuters.[18] This involves deciding on whether certain items such as editorials and news items should or should not enter into the denominator of the JIF calculation.

Manipulations, citation bias and gaming. There are numerous problems with use and abuse of citations.[19] These include self-citations, gender-biased citations, "rubber-stamping" highly cited papers and journals garnering increased exposure by placing accepted articles online[20] months before they are actually assigned to an issue.

Quality within. High impact factors and journal ranks can belie article quality, as measured by reproducibility and reliability.[18, 21] Low reproducibility can reflect underlying issues such as errors or misconduct.

Content inflation. Review articles are typically more cited than primary research articles.[22]

Trends in citation rate. Citation rates have generally increased across disciplines over the past several decades.[23] Thus, the average article published in the natural sciences in 1980 would have been cited twice in its first 2 years, whereas this figure was about five for papers published in 2015.

Differences between fields. Impact factors can differ considerably between research fields.[24] A notable attempt to normalize research article impact across disciplines is the Relative Citation Ratio,[25] which calculates citations of a paper divided by the expected number in the same field. Moreover, citations between different fields *within* journals are also expected to differ, meaning that the impact factor will be either an over- or under-estimate of the average for any given field.

Initiatives. Several high-profile declarations have fostered debate about the negative effects of citation metrics on the research community.[26] Notably, the San Francisco Declaration on Research Assessment (DORA) has as one of its three major strategic goals to "increase awareness of the need to develop credible alternatives to the inappropriate uses of metrics in research assessment." The Leiden Manifesto for Research Metrics is 10 principles for the employment of expert judgment based on actual research and transparency in the evaluation process. Finally, The Metric Tide examines the use and abuse of research assessment metrics by institutions and funders, and proposes "responsible metrics as a way of framing appropriate uses of quantitative indicators in the governance, management and assessment of research."

Breaking Out of the Numbers Game

The multitude of inherent biases associated with citation metrics and their use and misuse have been discussed for decades in publishing circles. Most of this literature was unknown to mainstream researchers until relatively recently and we see increasing discussion of issues across the disciplinary spectrum.

It is only with time that committees have become increasingly cognizant that citation metrics are not single number "magic wands" that completely describe the qualities of a scientist, article or journal. In this light, citation metrics will need to play a more qualified role in committee evaluations, and accompany more direct evaluations of scientific excellence, creativity, etc. found in a candidate's published articles themselves.[27] When metrics do enter into decisions, multiple rather than single metrics should be used, and their use needs to be tempered by their limitations and biases. A more reasoned use of citation metrics will mean that committee members won't save time and effort as hoped—it could rather be the opposite. Educators have a crucial role to play in training students, postdocs and young faculty who may be unaware of the many shortcomings and abuses in citation metrics.

24
Disposable Science

One of the great challenges to science is the ever-increasing number of publications. More items mean greater difficulty keeping up with the literature and judging novelty and relevance. One of the dangers is "disposable science"; that is, if an article is little cited in the first few years post-publication, then—regardless of its content—it may never have a substantial impact. Readers tending to cite already highly cited publications exacerbate the problem. This chapter discusses this issue and presents possible ways to reduce the loss of meritorious articles.

My doctoral research was on the interactions between natural enemies (parasites and pathogens) of insects. I used mathematical models and conducted controlled laboratory experiments and a field experiment. Part of the dissertation involved a complete literature review, which necessitated regular visits to the local library to retrieve and read relevant publications. I was interested in how interactions between the natural enemies changed through time and the effect this had on the host insect population. I studied numerous facets of the three-way interaction with the objective of understanding what made each population tick.

These were the early days of personal computers and I didn't have access to an electronic article database—I kept a card file. At the end of my dissertation, I counted the number of cards. They came to about 400. I had scoured the literature and, as far as I was concerned, I had found *each and every* published article or book chapter relevant to my area of study.

In the 30 years since my thesis, the number of published articles has increased several-fold. And that is not all. The number of articles published *per year* has also increased—growing at a rate of about 3.5 percent per annum over the first three centuries, and from 2001 to 2013 this increased to about 6 percent.[1,2] The total number of article records in the highly curated Web of Science is about 70 million.[2]

Science is cumulative and it accumulates.

> *Derek J. de Solla Price* is often credited as the "father of scientometrics." He wrote the seminal book *Little Science, Big Science … and Beyond* in 1963. Among his important insights was the prediction that scientific productivity would continue to grow exponentially, but there would be a slower increase in the number of quality researchers, meaning that an increasing proportion of science would be of low quality. He predicted that science was unsustainable and a "scientific doomsday" was less than a century away.

A primary research article nowadays typically cites more than 50 papers and it's not uncommon to see over 100. To arrive at such numbers, the authors would have likely read more than 200 useful articles. Many of these would not have been cited either because of the authors' view that it was unnecessary or limits imposed by journals.

Due to the sheer volume of published science, the majority of the articles we now read simply cannot be cited. Worse, given time constraints, many potentially relevant articles are not even read.

Science has become disposable.

Some Papers Emerge, Others Don't

More papers is more knowledge and more understanding. But acquiring this knowledge assumes that we can actually process it and zero in on relevant findings. This is a considerable challenge even for the most avid reader, and would be utterly impossible if it were not for three phenomena:

1. New findings often *languish and die*. If a new paper does not garner attention in its first months or years of life, then it is unlikely to ever be recognized as a scientific advance. This does not mean that the paper remains uncited, but rather

that there is no discernable catalytic effect resembling the black curve in Figure 24.1.

2. New findings that do emerge often *update or replace* previous results, rendering the latter less necessary to include on a reading list or to cite in a paper.

3. Scientists can modulate their preference to be more discerning in what they read. Some may still digest increasing numbers of articles, but it is an *ever-decreasing proportion* of what is actually being published.

> *The 80:20 law.* Citation rates in the mid-twentieth century were skewed such that 20 percent of published work received 80 percent of the citations. This ratio has shifted to 40 percent producing 80 percent of citations in 2005, possibly due to increases in the representation of journals publishing highly cited work.[3]

A scientific paper is metaphorically akin to a living entity. It is born at publication; matures as it is read, discussed and cited; may or may not continue to grow, produce "offspring" studies and possibly become a classic[5]; and except in rare circumstances (e.g. sleeping beauties, classics), eventually goes in decline (Figure 24.1).

Many papers however do not follow this path and are either rarely cited or never cited at all. Estimates of purely uncited papers vary considerably (from a few percent to over 50 percent), depending on the data source, time period, journal of publication and discipline.[6]

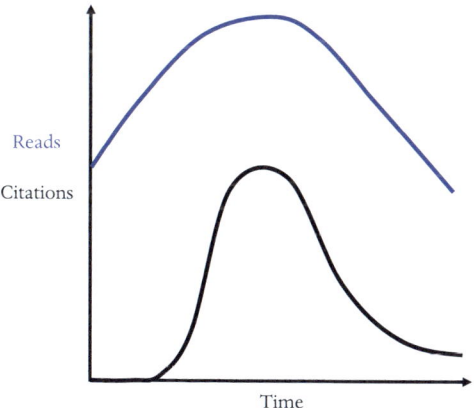

Figure 24.1 Hypothetical lifespan of a paper. Upon publication the article is gradually accessed, with maximum growth of the average article occurring in years 2 and 3, but with considerable differences between disciplines (see e.g. figure 11 in Rowlands et al).[4] With some time lag, the first citations appear and peak at some future time. Eventually, the reads and citations go into decline. However, because the literature is immortal, a largely ignored paper can be resurrected at some future time (a "sleeping beauty"; not shown).

Why do some papers dawdle in obscurity? The simple answer is that not all science is created equally. Whether a paper emerges depends on a variety of factors.

Finding science and limits to citing. Some science languishes because it is difficult to discover, locate or access. Database incompleteness is one reason; paywalls are another. Most journals impose limits to citation numbers, meaning that some references may not "make the cut" because, for example, they are not as topical as other (more citable) references.

The science. Some papers do not emerge because the science is simply not very good, not timely or too specialized.

Tested science. You have the choice between two papers to cite. The first was published 3 years ago and has garnered 50 citations. The second was published last month and has not had time to be cited. Which would you cite? If you are working directly in the area, then you will make a judgment based on scientific quality and relevance to the claim to be supported. But if you are a bit outside of the area you may be concerned that you cannot properly judge contents. You are most likely to cite the time-tested paper.

Names. Names influence why some papers (and authors) are more cited than others. Names include (i) highly cited, authoritative authors; (ii) high-impact, prestigious journals; and (iii) citing articles that cite *other* highly cited, prestigious authors or journals. It is perfectly reasonable to expect that—on average—highly cited, authoritative, reputable authors or journals do publish more meritorious research. But averages belie underlying distributions. Some papers by unknown authors, published in little-known journals, and not well cited may be higher quality, more interesting and more important science than others authored by big names, published in major journals and highly cited.

The Matthew effect, or accumulated advantage, is when one's early accomplishments beget status, credit and further accomplishment.[7] In other words, "the rich get richer and the poor get poorer." Examples of the types of achievements exhibiting the Matthew effect include: publishing in high impact journals, being cited, science funding, invites to present talks and academic prizes.

Cite what is cited. Subtly different from tested science and name effects is citation purely based on existing citations. Now, you are choosing between two papers published 3 years ago, where one has received 50 citations and the other five. Unless you read the papers carefully to decide which to cite, you may just go with the majority vote. This leads to "preferential attachment" and a skewing in citation distributions.[8] What is sinister is that many of the 50 citing articles may have cited the paper because it was previously cited! Indeed, this is also a reason why some papers *never* emerge: regardless of the merits of a paper, not receiving many citations in the first years of life could be taken to indicate that the study does not merit attention. The lack of citation is perpetuated by the lack of citation!

In sum, more than just content and relevance enter into citation decisions. Some papers get indefinitely pinned at few or no citations. Many of these would not emerge even if citation biases were to vanish, but some arguably would. Being passed over during the first crucial years can relegate a good paper to perpetual obscurity.

Possible Solutions

Citations are often subjective, influenced by opinions and open to bias. Because there is so much citable work and no regulatory authority, we are largely free to cite as we wish. The only real checks on this are having a method (see Chapter 3), our own integrity, and corrections and suggestions from reviewers and editors.

Author integrity. Author integrity is reflected in the science and how it is written. The scientific standard extends to all aspects of the article, including accurate and appropriate citations. High citation standards require understanding the literature and consistently applying norms in citation. All too often, papers cited are not read carefully or are cited based on "the front page" rather than the content. Conscientious citing is important in primary research articles, and absolutely crucial in review and synthesis papers. The latter two are places where both specialists and non-specialists are particularly likely to consult the cited literature.

Journals have a role to play. Journals rarely have much to say about citation policy. They assume that author integrity guides all aspects of a study including citations, and that the dedicated checks will come from reviewers. Although it is not the journal's place to suggest how to cite any more than how to do science, simply indicating on their "Advice to authors" and on the "Reviewer assessment form" that they view accurate and appropriate citations as part of high scientific standards, could go some way toward increasing citation integrity. More engaged journals could provide references or links to what they view as good citation practice.

Reviewer checks. While reviewers are supposed to check the veracity of a manuscript's analyses, most would consider careful citation checks to be too time-consuming and low priority. This is both because of the magnitude of the task and the fact that often there is no single "correct" citation. Some reviewers never comment on citations, whereas others only do so in key places. Sensitizing reviewers to the problems of citation bias would not only contribute to higher standards, but also influence their own citation behaviors.

Review papers. Reviews are an invaluable resource to keep pace with the literature. Depending on the theme, tens, hundreds or even thousands of relevant papers may be published each year. A review both brings this literature into focus and puts it into perspective. The authors therefore have some responsibility in presenting an accurate account of the subject together with key literature cited. The key here is "key"—the amount of relevant literature can be huge and to keep the bibliography manageable, some level of selectivity is necessary. To accomplish this, most journals impose citation limits, obliging authors to think carefully about the papers cited. Typically, this could mean that some of the most innovative articles do not appear, either because the authors miss them, or because they have been generally neglected by the scientific community—are under-cited—and therefore of lower priority or possibly even viewed as somehow flawed. One way forward is for authors to focus their review on recent advances (and indeed many journals encourage or require this), meaning that there is more room to explore and cite some of the gems that would otherwise be ignored.

Citation appendices. It is not uncommon that authors run up against journal-imposed citation limits. Some journals are very strict on limits and make no exceptions; others will consider arguments for exceeding the limit, and even then, the extra margin is usually thin. One way to address this perennial problem is for journals to establish citation appendices.[9] With a citation appendix, a journal can enforce lower numbers of citations in the main manuscript making the authors think harder about what appears in the front matter, but also providing the flexibility to include more complete citation lists in electronic appendices for interested readers. Citation appendices would be no different from other supplementary information: they provide key background information and can be with or without commentary. With the advent of hyperlinks, a symbol could be included wherever additional citations are made, which when clicked on, would go to the corresponding citations in the supplementary information.

PART VI
OPPORTUNITIES

Many of the issues seen in Part V relate to incentives for publishing more and in the highest-ranked journals. Career development however need not be unbridled careerism. Practicing the scientific craft responsibly provides enormous opportunities in research and career development and contributing to the scientific commons.

Chapter 25 provides a vade mecum of approaches and conduct that young researchers should consider in developing their careers. Key to achieving high-standard, enjoyable science is doing it with others, and Chapter 26 champions this approach and discusses the central importance of leadership and communication in successful collaborations. Chapter 27 argues that science goes well beyond original research articles and that young researchers—either alone or with more experienced colleagues—have the opportunity to make important contributions by writing reviews, opinions and perspectives.

Young scientists can also actively engage in the scientific community, possibly the most important way being peer review, and Chapter 28 presents how to write a useful report. Community also means communication, and a major innovation over the past decade has been the use of social media for interacting with others and exchanging information. Chapter 29 outlines this rapidly changing landscape, including its potential and pitfalls. Finally, engaging in the scientific community means enacting a sea change in the way young people learn how to write and publish their work. Chapter 30 closes the book with the main messages, a brief on where we are going, and an appeal to educate young scientists into the world of scientific communication.

25
Developing Your Career

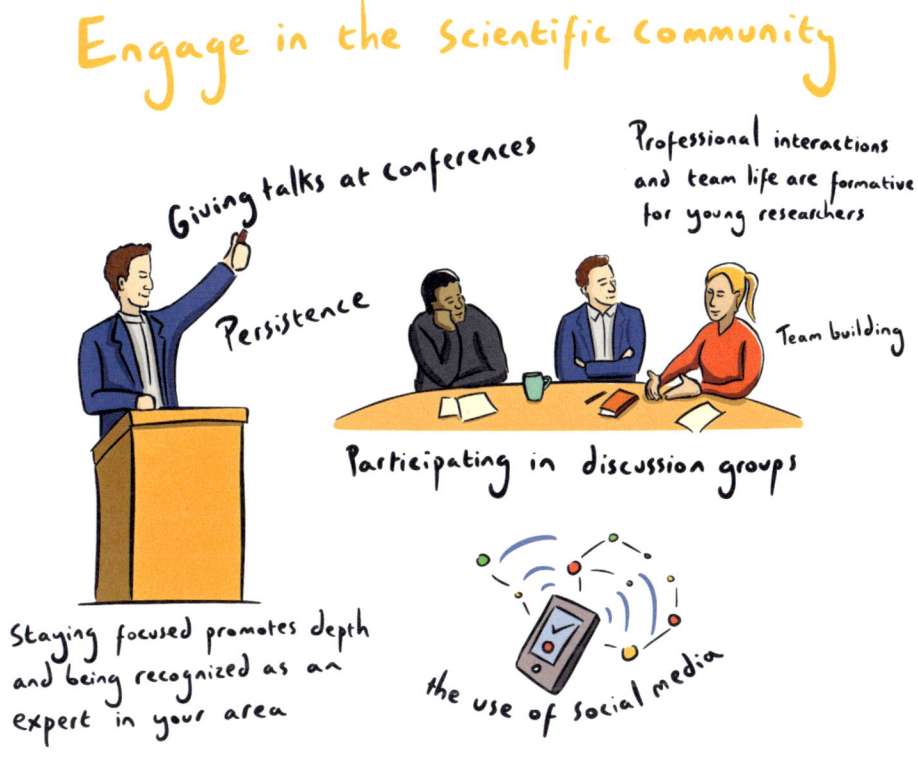

The publishing landscape is changing rapidly and in complex ways. Young researchers need to navigate this and find their place in the scientific community. Your place will evolve throughout your career, and your publication and communication strategies are keys to this path. This chapter provides some guidelines for how to embark on a successful scientific career.

Now that we near the end of the book you may be saying to yourself "Wow, with all this useful information about writing papers, publishing and communicating as a scientist, the author must have *breezed* through his career!"

Not so.

Much of what I've learned has been through my own mistakes and seeing the mistakes of others. Bad writing habits, submitting to the wrong journals, not communicating well with colleagues, getting involved in non-productive collaborations—just to name a few. We gradually discover how we work best as individuals, as part of a research group and as a member of a collaborative team. It isn't easy. Personalities enter into play, as do ambitions and constraints. Time fosters experience and—in getting better at what we do—we also become more self-confident. Self-confidence and growth are mutually reinforcing. We become experts.

Moving forward means being challenged. Some challenges are shared by all researchers, obvious ones being publishing that "big paper," finding that next research contract and learning new techniques that will take one's science to a higher level. Young researchers experience these, but also others that are more specific to early careers.[1]

This chapter presents essential ingredients for embarking on and navigating the complex world of a scientific career. Gaining expertise is a tall order. New challenges are many, diverse and sometimes unexpected.

Persistence Pays

Whether conducting a research project, publishing an article or applying for grants or jobs, you *will* encounter difficulties, setbacks and failures. Sometimes we agonize trying to figure out why, other times we understand the problem but are unable to resolve it. It's easy to get frustrated when things don't work and, to make things worse, everyone else appears to be succeeding. This is because you don't see their failures.

> *It's all about perspective.* No one celebrates a rejected paper. Innumerable hours went into the research and its write-up, and you had high hopes it would get published in the target journal. Now you need to revise and submit to another journal—a clear setback. But look at this another way. You approached your preferred journal knowing that there could be three basic outcomes: desk rejection, revisions and acceptance, or rejection following peer review. Desk rejection is irksome, but usually quick—with little effort you can approach another journal. Revisions are sometimes considerable work and reviewer comments can be annoying, but in "climbing this mountain" you are both improving your paper and planting your flag. Rejection after peer review is the worst of the three, but the compensation is extensive comment and critique that can be selectively used to revise and approach another journal.

All of this is part of being a scientist.[2] Easy science is likely to be boring science. Interesting science—almost by definition—*has* to be challenging, otherwise someone else is likely to have already done it! Interesting science is filled with unknowns, dead-ends and risks of failure. Metaphorically, each time you conduct a research project, establish a collaboration or write a manuscript, you climb a mountain. Often the climb turns out to be much harder than you had originally anticipated. This is why you need to go into the climb with enthusiasm and dedication. Enthusiasm and dedication give one the key ingredient to reach the top: persistence.

Persistence is an essential trait for scientists. Not enough means giving up easily and not making progress. The non-persister lacks scientific direction, and lives with frequent failure and Plan Bs. Too much persistence however means incessantly looking for why things don't work, spinning one's wheels and being victim of the "sunk cost fallacy" in Chapter 6. Both unbridled and insufficient persistence stymie progress and lead to anguish.

A young scientist needs to find the middle ground. This starts with establishing a research plan from the very start of a project. The plan will have a calendar indicating dates at which different milestones should be achieved. Progress needs to be monitored on a regular (but not too frequent) basis. Milestones should not be revised, but rather observations and contingencies should be added to the calendar as the project goes forward. Observations are important in keeping the researcher on track. Contingencies are critical in rebounding from setbacks, and if progress is really not going according to plan and options are few or exhausted, then a decision is made to move to another—usually previously discussed—project.

The decision to stop a project and move on is not as simple as it would appear. Tens and more likely hundreds of hours have gone into what is *your* project . . . and now the plug is pulled! Even if there is an equally exciting replacement, the sense of dashed hopes, wasted time and failure can be dispiriting.[3] This highlights the importance of discussion between young researchers and PIs (or more senior colleagues) in deciding how and when to abandon a project.

Active Engagement

Being a knowledgeable scientist obviously means gaining that knowledge. Imagine doing this passively; for example, when you happen to meet a colleague at a conference or by chance see an interesting article in the social media. These sources can contribute importantly to knowledge and insight, but depend largely on external events coming to you.

Although much of what you learn *will* make its way to you, a lot of it won't. Active work is necessary. Actively engaging in the same kinds of events—reading articles, discussing with colleagues, giving and hearing talks at conferences, and the use of social media—is key in determining research directions and becoming known. Active engagement is crucial both for your scientific growth and career development.

Finding the right balance between interacting with research and actually doing research is important. If you spend most of your time giving talks and attending interesting conferences and workshops, how could you possibly plan, execute and analyze experiments, write grant proposals, supervise students, teach courses, etc.? Some scientists *are* able to do both, but they are usually experienced and in high-functioning teams. The real risk for younger scientists is getting bitten by the conference bug, and not spending enough time at the bench.

Balancing means establishing priorities (experiments, teaching, manuscript writing) and reserving the remaining time slots for engagements. "Two hours every Thursday morning to read manuscripts, the 3rd to the 5th of September for an important conference, etc." Metaphorically, organizing one's schedule between priorities and interesting but sometimes unnecessary events is like playing the video game "Tetris." Facultative blocks are oriented into the spaces left by unmovable essential blocks. Planning, organization and opportunism are all important qualities in balancing and making the most of active engagement.

> *Job advice.* Attend conferences, workshops and summer schools. PhD opportunities, postdoc offers and even faculty positions can originate when people actually meet you and hear about your research. PhD and postdoc opportunities in particular are often announced at the end of research talks.

Staying Focused

Perhaps the biggest mistake that a student can make is to diversify too much, too early. A diversity of scientific interests *is* important for scientific development, but bringing *too* much into your own research means less depth in each and fewer significant contributions. There are two main issues here.

The first is that in the initial phases of a scientific career you are gaining expertise. This takes time and, as the saying goes, time is limited. If it takes you 3 years to attain a given (high) standard on your dissertation or postdoctoral project, then chances are that you would be hard-pressed to achieve anything like that if you were to conduct two different projects. This said, an interesting side project can be a learning experience and a "credential builder." Successful side projects are often collaborations among graduate students and postdocs within a group, with a small and well-defined question that can be answered in a short time.

> *10,000 hours.* In a famous article in *The Scientific American*, Herbert Simon and William Chase[4] estimated that about 10,000 hours of dedicated work was necessary to become a chess expert (approximately the top 5 percent of rated players). Malcom Gladwell developed this idea in his book *Outliers: A Story of Success*. In a nutshell, expertise takes time—a lot of time. Of course, some achieve a given level faster than others, but the basic principle still holds. When endeavors require different expertise (e.g. sprint vs. marathon runner), investing in one area will mean sacrificing it in other areas.

The second issue is how the scientific community perceives diversity. Seeing a name as lead author on papers treating very different subjects can invite skepticism. "How could that person have mastered such different topics?" Certainly possible, but some will still believe that had the scientist just focused on one of the subjects then she would have done deeper, higher-quality research. Potential employers may see risk in a young researcher who appears to frequently dabble. Diversity *is* good, but so too is having a central, identifiable question or theme.

Careful Choice of Journals

Choosing a journal was discussed in Chapter 16, but I can't emphasize enough how much journal choice *can* make a difference and so revisit it here.[5] Aiming too high can be a waste of time, ending in rejection and frustration. Aiming *too low* in a very specialized or regional journal can mean your article is lost in oblivion. Avoid a hasty or uncareful choice that you come to regret. Once the paper is published, you can't go back and withdraw it and submit it elsewhere.

We often don't take enough time in choosing a journal—it can be tedious, involve protracted discussion and lead to arguments. Choice is frustrated by the many unknowns in how different journals operate, what the likely trajectory of your manuscript will be and how your paper—if published—will be received by the scientific community.

Choice can also be affected by enthusiasm and time. We are more likely to carefully study candidate journals for the first submission than when the paper has been rejected several times and a year has gone by. Clearly, time elapsed is an important factor, but it should influence choice, not rush it. Hasty journal choice due to exasperation is something that one may come to regret.

Journal choice needs to be methodical, yet adaptive. It is not as simple as establishing an immutable ranked list from the beginning. For example, should your paper be rejected without review by the top journal on your list, then maybe you will think twice about submitting it to number 2 on your list, for fear of the same fate. Should your paper be peer reviewed but rejected by the new number 2, your original choice for number 3 may change based on reviewer feedback. Establish a list from the start, and expect it to change.

Careful Manuscript Preparation

I've belabored the importance of clear, concise and accurate writing throughout this book. There is little point in publishing a scientific paper if readers can't follow it or, worse, accept errors and misunderstandings as fact. Good writing respects the science and fosters your reputation as a scientist.

Quality writing is not only about weaving an interesting story with an enticing puzzle and peppered emphasis. It's about accuracy, clarity and precision. It's about avoiding errors and ambiguity. Errors accrue when one writes quickly, sloppily and does not check (and have co-authors verify) facts, be they numbers, equations, claims, etc. Errors that make it through peer review and are published may never be discovered by the authors themselves, and, when they are, can only be corrected via an Erratum, which, although scientifically necessary, is also an announcement to readers that the authors may have been careless.

High-quality writing requires meticulousness. Ambiguous or inaccurate sentencing and terminology can lead to misinterpretation. Readers who don't understand or misunderstand parts of a study may either discount the article's importance (and ignore it) or come away with mistaken inferences. Unlike an actual error in a number or an equation, inaccurate, unclear writing usually cannot be corrected through an Erratum. Achieving high-quality writing is evidently challenging for those who are new to written scientific English. Different options are available, including soliciting professional editors or editing services at your own institute, and contacting an English-speaking colleague who would be willing to be involved in the write-up of your study.

Writing is not the only factor in careful manuscript preparation. Figures are particularly important. A good figure presents information or results that could not be easily expressed in words. A good figure also catches the eye. It is both aesthetic and interesting, yet is not cluttered, has easily distinguishable symbols and lines, clearly labeled axes, etc. A good figure can be (almost) understood without an accompanying figure legend, but a good legend is necessary to make the figure as self-contained as possible.

Professional Relationships

Good communication channels make good, enjoyable research. Poor communication can lead to misunderstandings, arguments and in some situations, falling-out.

Effective communication needs chemistry. It should come as little surprise that the most enduring associations have one or more of: personal affinities, complementary scientific backgrounds or a track record of publication successes. Sometimes we hit it off from the start, with others it takes time. But often the chemistry is simply not there—communication is cumbersome and even annoying—either projects don't take hold, or they do but the road is rough. Communication issues are certainly not a reason *not* to collaborate, and some of the most exciting research emerges from challenging interdisciplinary, international and multi-technique associations.

Evidently collaboration is a major form of scientific association, but positive interactions may also involve advice and assistance, such as reading and commenting on manuscripts and providing letters of recommendation. Forging professional relationships and culturing long-term ones can be an important part of a young researcher's scientific and career development.

Building the Team Like a Family

Research teams are in many respects like families. The structure may be anywhere from hierarchal to egalitarian, but almost invariably has only one responsible person—the PI. As a student, postdoc or young faculty member, you will have experienced one or more teams. There is no tried and true recipe for organizing a team; rather, the PI will probably import part of what will become the team's culture from her own previous team experiences.

Fun, insightful and productive research will depend on personal and professional support among team members and—for teams of a certain size—the more experienced students and postdocs helping or co-supervising the less experienced. Fitting into such a structure is not always easy. Most new students and postdocs learn on the job through interactions and team meetings.

Teams, like families, experience both happy events and challenges: members joining the team, leaving the team, publications or grants accepted or rejected. A healthy team has an identity and a culture, but nevertheless evolves scientifically and socially.

A young scientist who is appointed to a faculty position will encounter the onerous and scary situation of setting up her new team. Typically, colleagues and faculty members at the host institute will help the new faculty member get settled in. From there, she will begin building her team, focusing on achieving milestones such as equipping a laboratory, attracting a group of students, submitting research grants, completing previous studies and embarking on new ones.

Establishing a research group takes considerable time and energy. The new faculty member will probably invest most of her time in working to establish a good environment and vigorous research program, and considerably less in hands-on research. To facilitate this process, the host institute usually reduces or waives teaching and administration responsibilities for a negotiated period of time.

Contribution to the Commons

Occasional helpful behaviors can have major positive effects on a group and beyond. This works for collaborative groups, research teams, university departments, academic societies, and extends to functions such as committee work, peer review and acting as editor, teaching and interfacing with the public.

It is not always straightforward working out how to establish a balance between the demands of various extra-research activities and scientific research. One may find oneself either required or compelled to contribute to departmental or scientific communities—teaching and committee work in particular—and sometimes the task can be enormously demanding. Whether required or voluntary, rigorous committee work is not only important, it earns the respect of peers. Unfortunately, many of our activities—such as anonymous grant and journal manuscript review—do not gain the recognition that they deserve, but the latter is gradually changing with Open Science.

My advice to students and postdocs is to embrace responsibility, but learn when and how to say no. Typical responsibilities include teaching, grading papers and laboratory assignments, contributing to the smooth running of laboratories and giving guidance to less experienced team members. Often young researchers do not know how much they should contribute and what is really useful—depending on the context, they should not hesitate to talk to peers, more senior colleagues or PIs.

26
Collaborating

Collaborations create opportunities to learn from others and conduct higher-level science. But they also present significant challenges, including interacting as part of a group and your role in contributing necessary elements to a study and its write-up. This chapter discusses the skills that foster a great collaboration.

Working solo has its advantages. You make your own schedule, decide what you want to do and how to do it. You get full credit. If you are single minded and determined to do something in a particular way, then going solo may be your best option.

I can speak from experience. Many of my first papers in the 1980s and early 1990s were solo efforts. This reflected the times and norms in my scientific area. It was totally possible and even well looked upon for a researcher to conduct a study solo from beginning to end, especially for papers resulting from a dissertation. The basic steps were the same as they are today: having read the literature I would be intrigued by an unsolved problem; develop the main question into a series of more specific questions; determine the scope of my study and the experimental approach; execute the experiments, analyze the data, generate tables and figures, write up the manuscript; ask close colleagues to read and comment on my paper; submit to a journal of choice and deal with eventual rejection or the possibility to resubmit, which entailed revising the manuscript and convincing reviewers and the editor that my manuscript now merited publication.

Except in rare cases (such as an invitation to write a perspective article), I no longer solo-author publications. The reason is simple. Good, interesting science is much more complex nowadays than it was 30 years ago. A single person would be at great pains to produce a study that has the quality and scope of one born from a collaboration.

Collaboration is a mainstay of science.

A Simple Scenario

You are having coffee with a colleague and discussing some recent research. The conversation is fast, exciting, takes abrupt turns, increases in depth, volume, has frequent toing-and-froing and draws a sweat. You pause a second, realizing that a watershed of highly relevant primary research papers has been published recently. You both discuss the main features of this work and one of you blurts: "It would be absolutely GREAT to review this!" The other immediately replies: "That's *exactly* what I was thinking. LET'S DO IT!!"

Indeed, many collaborative studies actually begin like this: the generation of a great idea and realization that you can plant your flag together. As long as you can keep the momentum going, milestones emerge and eventually the mountain peak comes within sight.

Such eureka moments sometimes do bear fruit, and may even produce a *better* paper than one emerging from the protracted search for an interesting problem. But spontaneity also has risks. The unbridled enthusiasm in the days and weeks following the moment can blind one to the possibility that a study may not reach expectations, or never even be completed. Worse, you may fall victim to the "sunk cost fallacy," where long after the initial euphoria and the divvying up of tasks, you find yourselves with a study that is not even coming close to your original aspirations. And now that you have invested *so much* in the endeavor, you must continue until it's completed! So, you continue in the hope that things will improve and eventually produce a good study—or at least publishable one.

There is always a risk that a project does not produce a paper. Even the most carefully planned studies sometimes do not bear fruit or they produce less impressive papers than originally expected. But risks can be managed. Keep up the momentum and the spark until you are ready to discuss the *serious* work involved. It may be then or sometime after that the decision will be made whether the project can really take off.

The above reality check is not easy. Different scientists pursue projects with different levels of zeal and investment, and have contrasting professional obligations and ways of prioritizing. This leads to the frequent situation where projects are delayed or simply dropped. Although disappointing, you will have learned, hopefully strengthened your camaraderie with your collaborator(s) and may indeed revisit the project at some future time.

The Team

As a solo author, what we gain in autonomy and pride, we may lose in quality, scope and depth. It's straightforward to see why. Collaborators bring their complementary masteries into a single endeavor. The whole is often greater than the sum of its parts and far better than any team member could have possibly accomplished alone.[1] Metaphorically, conducting a study as a group is akin to building a house: you need specialists—masons, electricians, plumbers—but also a generalist—the foreman—to coordinate and verify the work.

> *Is there a leader in the house?* All too often collaborations can stall or stop because of leadership issues. Projects typically run smoother when a single person (such as a PI) takes on the lead role. But multiple leaders, particularly colleagues who have some prior experience working together, can also productively direct a project, especially for large and diverse collaborative teams. Leaders organize the collaborative effort, motivate the troops, monitor progress and oversee the scientific quality of the project. Good leadership requires personality, organization and patience.

More practically, if you consider your own abilities relative to peers, you will undoubtedly see that you are the strongest at some capacities, like experimental design or literature review, and possibly weaker at others, such as generating figures, clear concise writing, etc. How does this actually play out in a collaboration?

Usually, collaborators will divvy up tasks according to who wants to do what, and who is best positioned to do what. There will invariably be overlap in abilities. Not only is some overlap good for a study, sometimes it is necessary. Imagine an extreme situation where everyone does her or his own part but has no expertise to usefully comment on the contributions of others. Flimsy logic, errors and imprecise, ambiguous writing would remain uncorrected. Once submitted, such a paper would be at the mercy of editors and reviewers.

> *The polyvalent specialist.* Expertise in a particular area is important both to attain professional standards and to be a useful part of a collaborative team. However, to have completed your doctoral dissertation, you will have necessarily acquired some level of polyvalence. This includes an in-depth understanding of your system, scientific methods (such as experimentation), data analysis, literature review and scientific writing. Polyvalence should be nourished both for being a broad-thinking scientist and for the spectrum of your contributions to collaborations.

Overlapping contributions, effective oversight and good communication are essential ingredients for a high-quality, seamlessly flowing manuscript. Indeed, a sense of "dedication to the team" potentiates individual contributions: one wants to achieve more and better *for* the team. We learn a great deal in the process not only because we strive to do our own part as best we can, but because we need to interact collectively to achieve a common goal.

When teams work well they can produce work that could not be otherwise attained in terms of scope, quality and achievement. However, collaborations are not static—logistics, perception of the project and priorities change. Sometimes collaborations nevertheless lose their "freshness." The project and people's roles are established and completing the project is only a question of fulfilling one's responsibilities. One of the dangers of routine is not daring to discuss aspects of the work that appear to be "written in stone," but are actually amenable to change. Obviously, some parts of a study are unchangeable once engaged, such as an experimental design, but others, such as formulating new experiments or proposing new/different figures for a manuscript, can make a huge difference to the final publication.

Teamwork extends well beyond scientific method: it includes work ethic, approaches to manuscript writing, perspectives on journal choice, and turning a published study into seminars and media exposure. Working as a team is particularly beneficial to younger scientists, since they are in an ideal environment to learn from more senior colleagues. When we learn from others we not only have first-hand access to recipes, but also to the teacher's *experience*—the small "tricks of the trade." Even should a more senior colleague be chronically busy, minutes or hours of interaction likely will transmit insights that would otherwise require months or years.

> *Pros.* There are numerous reasons why collaborating is better than going at it alone. These include greater/higher/better: quality, impact, scope, learning and productivity, and to achieve studies that would otherwise not be possible. Collaboration also fosters the joint control of quality and of cooperation rather than competition between research groups.

Trust

Working as part of a research team requires cooperation, coordination and trust. The first meeting with potential collaborators can be crucial not only to see whether a mutual effort will emerge, but if it does, to give an idea of how different members will interact and contribute to the project. Some of us are particularly good at rapidly inferring affinities. But rare is the person who can really predict the future dynamics of a budding collaboration. Rather, many meetings are usually necessary to define roles, responsibilities and eventually authorship. This hints at the complexity of how meetings influence a collaboration. The dynamics of a meeting will depend on its organization, who leads the discussion, key moments and decisions, candor, mood, etc.

Building a mutually trustworthy collaboration can take time. Participants want to see that everyone else takes the collaboration seriously, which means, for instance, carefully preparing for meetings, doing concerted work on the study and contributing to progress reports. Trust emerges when everyone sees that everyone else is contributing.

> *The Ringlemann effect* comes from management science and refers to the tendency for each member of a group to expend less effort as group size increases.[2] In scientific collaborations, this can manifest as delaying or reducing contributions and looking to others with similar expertise to do the work. The Ringlemann effect can thus lower quality and productivity. This phenomenon can be minimized through both good communication and focusing on project goals from the very start.

Formality, personality and chemistry are all important in collaborations. Be yourself, but be conscious that over-informality may not be to the taste of everyone in the room. As a rule of thumb, it's good to err on the side of reserve, and dose any informality according to the behaviors of others in the group. This may sound overly calculating, but some working associations naturally invite personality, whereas others (particularly committee work) are impersonal and serious. One usually sees in the first minutes where on the formality–informality spectrum interactions will lie.

Non-Starters, One-Night Stands and Lifetime Vows

No two collaborations are alike. Metaphorically, some are lifetime vows, others one-night stands and yet others never get off the ground. Whereas it is usually easy to predict a non-starter, it is hard to forecast how long a successful collaboration will last. Not only are there factors such as age, academic background and experience, goals and personal affinities, but also more complex parameters such as the number of collaborators involved and their individual commitment.

The division of labor is particularly important and a good predictor of the success of any project. Who does what and by when should always be established early, monitored and reviewed. It is equally important as mentioned above that one or more persons has enthusiastically accepted leadership responsibilities.

Most successful associations produce only one or a few papers. Some are intentionally one-offs (e.g. a paper emerging from a scientific meeting) with no intention to reconvene the group. Others are one-offs because not everyone was happy or because some collaborators simply decide to move on to other projects. Some associations never produce a paper, either because projects did not bear fruit, or because the association never intended to produce publications.

Occasionally, what started as a one-off continues to flourish and becomes a long-term association. Career-long collaborations are rarely foreseeable, but when they do happen can produce an important body of work and be very enriching for those involved. Some collaborations endure because of a good match between the people and complementary scientific approaches and goals. Many long-term associations proceed one paper at a time; others are at the center of each participant's research program.

Meeting and Workplan Basics

One oft-neglected element of collaboration success is organization, and in particular, creating an enjoyable and productive meeting environment. Below are some of the basics to consider in organizing a collaborative group meeting:

Logistics. Participant travel and lodging.

Amenities. Comfortable meeting room isolated from disturbance; video projector; whiteboard; refreshments; group meals; and a team outing.

The meeting and work plan. Provide agenda (presentations, discussion and work periods); set goals and agree on questions and methodological approaches; discuss who does what; coordinate software and word processing programs; and discuss next steps (individual assignments, video conference updates, next meeting).

27
Writing Reviews, Opinions and Commentaries

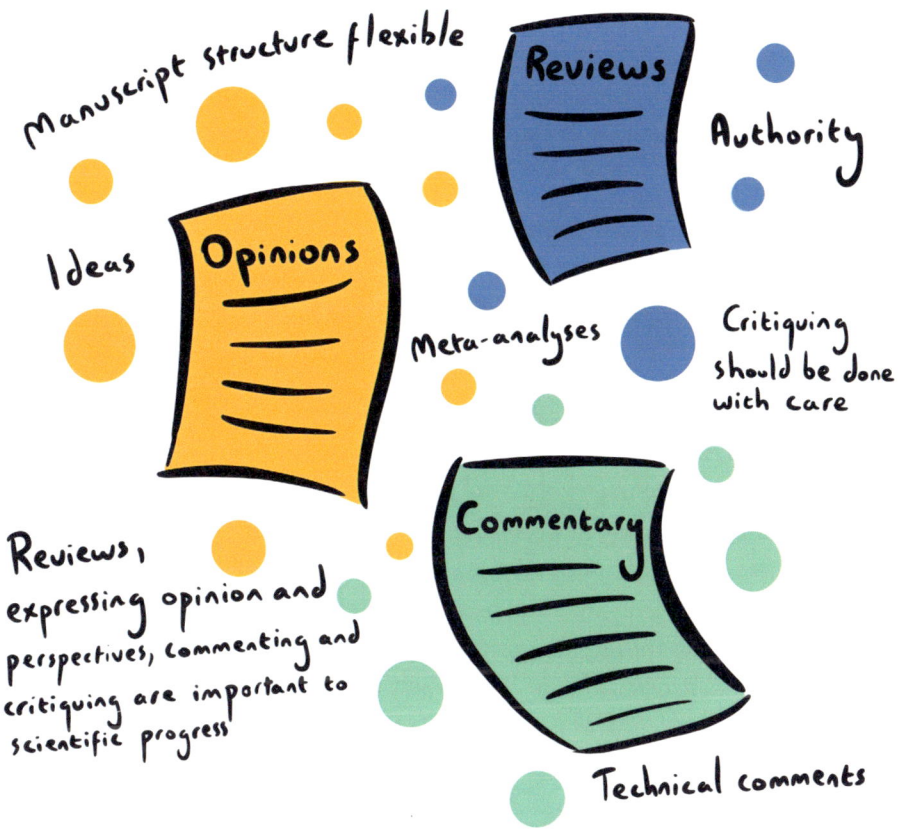

Scientific journals have greatly diversified their content over the past few decades. Forums commonly seen nowadays include: essays, opinions, perspectives, commentaries, news and reviews. Although not as technical as original research articles, these rubrics nevertheless maintain the essential pillars of accurate, neutral, clear and precise writing. This chapter presents several of the most known article types and the opportunities they present.

My first experiences in science were in the 1980s. Most journals (almost) exclusively published original research with just a spattering of reviews, opinions and commentaries. One went to journals like *Nature*, *Science* and *The Scientific American* to learn about areas of science other than their own. *Annual Reviews* would publish a collection of timely disciplinary syntheses. Disciplinary original research journals gradually saw the need to cover other aspects of science, but it was only with the *Trends* journals that everything really changed. In my area of research *Trends in Ecology and Evolution* launched in 1986, presented news, events, job adverts, opinions and reviews. Like most paradigm shifts it took time for scientists to integrate non-primary research into their scientific regimen, and for primary research journals to follow the lead.

Fast-forward to the present. Virtually all journals now have sections of non-primary research. Non-primary research papers (NPRPs) can have substantial impacts on a field. These range from synthesizing findings in a fast-moving area to challenging conventional wisdom. Rubrics come in all shapes and sizes. Simplifying somewhat, they usually either take stock—*reviews, syntheses, meta-analyses* (R/S/M) or present views—*ideas, perspectives, opinions* (I/P/O).

> *Reviews* and *syntheses* give the reader an in-depth presentation of the current state of a field (historical scientific progress in the form of a literature review), and structure what may be a large number of research themes into a few main avenues. Whereas a review generally doesn't present new analyses, syntheses often do. Many journals request that authors focus on recent progress (e.g. past 5 years). Sometimes the terms "review" and "synthesis" are combined into a single rubric, but even when not, they each generally cover the other to some extent. Some journals request that authors submit a proposal for approval before submitting a manuscript.

> *Meta-analyses* are studies of multiple datasets that test the generality of findings. They can be used to evaluate existing or new hypotheses, and aim to produce new insights about long-standing scientific questions or debates. Papers employing meta-analyses are in some respects hybrids between original research and review/synthesis articles. Should a target journal not mention or not have a specific rubric for meta-analyses, it is best to write to the editor to see how the journal handles such studies.

> *Ideas, perspectives* and *opinions* are usually developments that one would find in the Discussion section in an original research paper. I/P/O will (at least briefly) review the topic and present the background that leads to the reason for the exposé. There can be considerable overlap between these three rubrics and sometimes they are combined in pairs. Whereas ideas stress novelty, perspectives and opinions seek taking a stand on a topic. The science of I/P/O is the logical arguments, facts and associated citations supporting or refuting previously published or new claims. Similar to R/S, many journals require a proposal in deciding whether to encourage a submission.

Achieving the depth and scope required for an NPRP can be a highly enriching experience for young authors and adds significantly to their expertise on a theme or question. NPRPs are greatly appreciated by readers, who seek the insights that otherwise would be difficult if not impossible to obtain. Indeed, some NPRPs are extremely influential and among the most highly cited articles in their area.

Embarking

The impetus to write an NPRP can stem from diverse sources including a dedicated reading of the literature, a specific paper or achieving a milestone in one's own research program.

Consider the following realistic scenarios.

Scenario 1. In writing the Introduction of a collaborative original research paper, you have a eureka moment. You realize that despite numerous recent developments, no *review* paper has been published on your study topic for the past several years. The area is moving fast and in many directions. You believe that experts are increasingly challenged to navigate the area, and non-experts could be interested in getting a bird's eye view. You and your colleagues have already done some of the groundwork and are in the perfect position to embark on a *review* paper.

Scenario 2. An article in your area has caught your attention. You believe that the paper is important and that the majority of interested scientists may miss its relevance. You have the motivation and expertise to write a short commentary, and decide to contact the journal editor to see if this would be possible. Should the editor not be interested, other journals could be. *News* and *commentaries* are some of the most pleasurable articles to write, and carry with them responsibility to reach out to both specialists and the general scientific community.

Scenario 3. You are particularly critical of an article that's just been published and begin to assemble scientifically credible arguments and reanalysis that will test the paper's findings and claims. Not all journals accept targeted critique of their published material, but those that do will have a *technical comments* rubric or similar. Journals publishing *technical comments* usually give critiqued authors the opportunity to reply.

Not all journals publish NPRPs or, when they do, have a limited spectrum of rubrics. This means that some shopping around will probably be necessary. As mentioned above, many journals will not entertain spontaneous submissions; rather you will need to submit a carefully written proposal. Journals generally provide guidelines for what NPRPs are expected to achieve, and the same rubric (e.g. *reviews*) may not be identical in form at different journals. NPRPs are subject to editorial screening and usually also peer review.

Experience and Authority

NPRPs are advanced undertakings and successfully writing them comes with experience.

First, unlike original research papers, which generally follow IMRaD or a similar format, NPRPs have no set structure beyond perhaps what the target journal suggests. Journal guidelines can be helpful and are important, but the best way to decide how to structure your NPRP is to look at examples (i.e. *Models*) actually published in the target journal. Despite greater liberty in the narrative, NPRPs follow the same basic quality standards as original research articles.

Second, NPRPs marshal the writer's breadth and depth of knowledge. The prospective author needs to know the relevant literature, understand how concepts interlock, and have an overarching vision of the area and where it is going. Authors need to determine the key articles to discuss and cite. Often a young author will embark on her first NPRPs by teaming up with more experienced colleagues. Indeed, NPRPs are challenging for any single person to write, and greatly benefit from multi-author collaborations.

Although many highly influential NPRPs are written by experienced, authoritative scientists, writing a good NPRP can actually establish one as an authority. If you are a beginner and not yet confident about going about it on your own, as per above, consider associating either with a researcher in the same career phase as yourself, or with a more experienced colleague.

Etiquette

No one enjoys having their work critiqued. As thoughtful as one may be in writing a *commentary* or *technical comment*, some authors will still feel attacked. The risk of a critique being poorly perceived extends to readers too, who above all appreciate veracity and importance, and dislike flimsy or petty attacks. Nevertheless, a scientific writer cannot possibly anticipate all sensitivities, and presenting important hard facts should not be subjugated for fears of possible reactions. This can be done—without irking authors and creating a spectacle for readers—through careful writing. Avoid unnecessary expletives and exaggeration. Personal critiques are unacceptable. Have friendly reviewers read your manuscript before finalizing and submitting it.

Unique Opportunity

Some scientists never write NPRPs; others make a career out of them. You are likely to be somewhere in-between, writing or contributing to one every few years.

Unlike original research articles, NPRPs are almost exclusively one-offs—unique undertakings. More so than original research articles, their relevance can be time-sensitive. Clearly, a commentary or news article needs to be published either along with or shortly after the target article. The relevance and impact of reviews, syntheses and opinions too can be influenced by the current state of knowledge and understanding in an area. Publishing such a paper too early could mean not having a sufficient foundation to build upon; waiting too long to write a review risks being scooped by the publication of a review from another authoritative group.

28
Reviewing Manuscripts

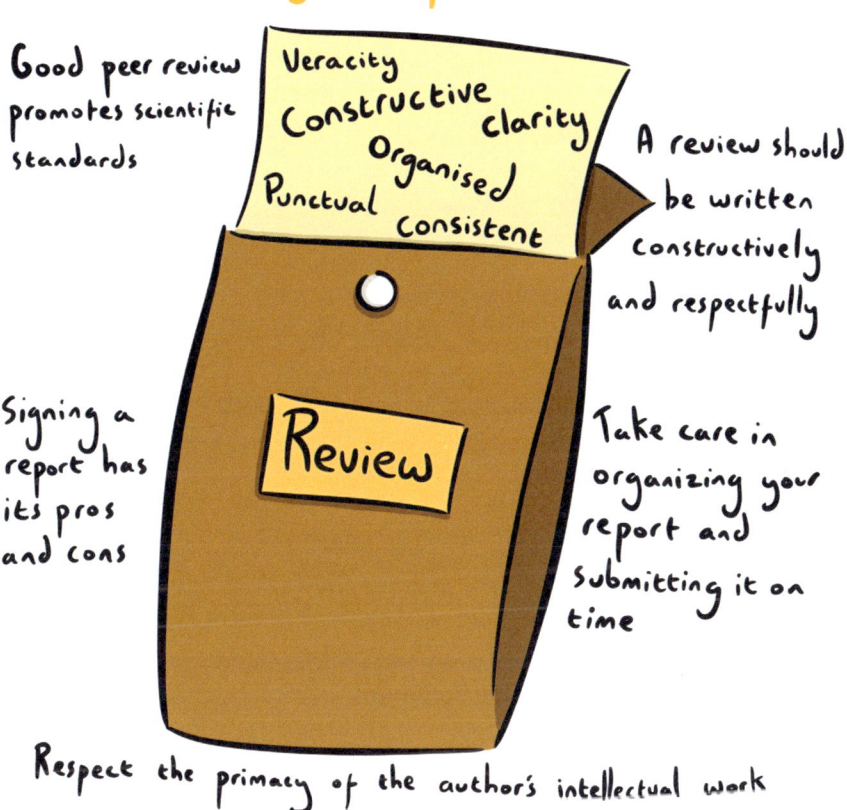

Peer review is a pillar in scientific progress. Scientists-as-reviewers assess manuscripts of scientists-as-authors with the central aim of improving quality. Contribution to peer review is voluntary and typically solicited only once one begins to publish. Given the central importance of peer review to science, it is unfortunate that it is often learned on the job. This chapter presents the basics and the important features of a useful review.

Reviewing manuscripts is the most fundamental contribution that a researcher can make for science aside from doing science itself. Conducting thoughtful and thorough reviews is a responsibility, but one that—as discussed in Chapter 21—is increasingly ignored or discounted. Reviewing needs to be revitalized as an *opportunity* to impact science through the identification of shortcomings, errors and flaws, suggestions for improvement and advice to editors as to whether or not a revised study could meet the standards of the journal.

> *Science evolves.* Science seeks to explain observation. Scientific understanding evolves when we test existing hypotheses and theories and obtain new insights. But the potential for human error—analogous to deleterious mutations in evolutionary biology—is ever-present in scientific study. Although far from perfect, peer review works to either remove (study is rejected) or correct (revisions) these. Science therefore progresses through discovery and review. In the absence of peer review, readers are the only stopgap. Should readers replicate these errors in their own work and the errors propagate, then science suffers.

Few realize just how important external, independent expert reviews are to the scientific process.[1]

Imagine a world without peer review: no comments on your manuscript beyond those made by your co-authors and, possibly, one or two friendly colleagues. Co-author and colleague input are essential, but limited for two main reasons:

- Co-authors are so close to a study that they simply fail to see many of its problems. This can be an issue particularly when each co-author has a narrow spectrum of expertise.
- Co-authors and colleagues do not want to ruffle feathers. Critiques—no matter how diplomatically expressed—can be challenging, embarrassing and even offensive to the scientists leading the study.

Friendly, non-anonymous reviews improve manuscript quality, but risk missing or avoiding more consequential problems that peer review is intended to identify.

Peer review aims to resolve ambiguities in writing, correct errors of method, logic or claims, and suggest general points of improvement. Errors present at publication will lead some perceptive readers to dismiss the study; other readers will fail to see the shortcomings and accept erroneous parts of a study as fact.

Peer review is not perfect. The two or three experts typically providing comments are an incomplete sample of the scientific community, and many peer reviews do little service or even disservice to authors. Nonetheless, peer review is currently the best means we have to maintain scientific standards and in so doing, foster scientific progress.

Peer review is rarely taught in graduate school curriculum and many PIs never discuss reviewing with students and postdocs. Most of us are self-taught peer reviewers.

Reviewers Need Help to Help

You receive an email to review a manuscript for a journal at which you have recently co-authored a paper. Of course, you are honored and excited that the journal considers you an expert! But you have never actually reviewed a paper before—the closest was to have commented on a manuscript by another doctoral student in your research group.

How much time should you spend reviewing the paper? What exactly does the journal expect from you? Should you sign your review?

Think of the courtroom analogy presented in Chapter 13. You are accustomed to being an author, that is bringing your case before the judge (chief editor) and jury (reviewers). You know that several weeks to several months after submitting your case (manuscript) to a court (journal) you will receive the judgment along with observations, comments and critiques made by the members of the jury. You recall how you tried to guess the identity of reviewer 1, who provided very useful comments, and wondered what could have possibly gone through the mind of reviewer 2, who clearly misunderstood your results.

Now *you* are a jury member. Your responsibilities are outlined in the invitation letter. You probably see the authors' identities on the title page, and realize that unless you sign your report, they will not know yours. In a cursory examination of the manuscript you see that it is far from its potential—you have a lot of work ahead. You try not to be influenced by what you would have expected to be a more finished product. You estimate how much time you can spend on the paper, look at the clock and start to read and take notes. Two hours later your report is done; it took more time than expected, but you have identified a number of shortcomings and even some scientific flaws. You feel satisfied since the recommended revisions should improve the manuscript and correct several (embarrassing) errors. You also have complimentary comments for the authors and begin to write your report on this positive note.

Although you do not believe that the paper's sometimes ambiguous writing influenced your understanding of the science, it might have for a less tolerant reviewer. You decide that it is important to inform both the authors (in your report) and the editor (in your confidential comments) that you did struggle with parts of the manuscript. You complete your report and advice to the editor and, beyond the scientific comments, state that writing improvements would be necessary in any revision.

The above generic reflects the mindset of reviewing manuscripts: the willingness to spend the time and effort to help authors improve their study, and to advise the journal whether or not to permit a revision. In accepting to review, one tacitly expects to comment mostly on the science and not be lumbered by the writing and presentation.

How long does it take? According to a 2015 PRC survey,[2] reviewers spend 5 hours (median; 8.4 hours mean) reviewing each manuscript, and the typical scientist reviews eight papers a year. The time to submit a report differed considerably between countries and disciplines, with a median time of 16.2 days.[3]

This is not always the case and, as pointed out below, why it is important not to review a manuscript that you do not sufficiently understand because of the writing.

Etiquette

There is no official etiquette in reviewing. Some of us learn from our supervisors, others from reading reviews of their own papers or of other papers, and many just use common sense. Most journals do list their expectations either on their website or more directly on the review form. These suggestions—when given—are often of limited use or simply ignored by time-strapped reviewers.

Below I have assembled the main points to consider in deciding whether to accept a reviewing assignment and if so, then your responsibilities as a reviewer.

> *In your area of expertise?* Don't expect that each and every time you are asked to review a manuscript, it will fall smack in the center of your own area of expertise. The two main reasons for being asked to review papers somewhat outside of your main area are: (i) the editor is actually looking for such views; or (ii) the editor could not obtain agreements from preferred experts and proceeded down the invitation list until reaching your name. If in reading the title and abstract in the invite email you believe that the paper is likely to be too far outside your area, then it is best to decline to review. Occasionally, however, it only becomes clear that you cannot confidently review the paper once you actually inspect the manuscript itself.[4] In such cases you should immediately contact the editorial office either to see whether you can conduct a partial review on parts of the study on which you feel qualified, or to inform the journal that you cannot provide a review.
>
> *Avoid conflicts of interest (COIs).* Journals have COI policies. COIs include, for example, publishing recently with any of the authors, or personal or professional associations or conflicts with any of the authors. Although not technically a COI, some reviewers would recuse themselves if they have previously reviewed the same (rejected) manuscript for another journal. Nevertheless, you may feel that despite a possible COI, you can fairly review the manuscript. In such a case you should send an email to the editorial office explaining your concern and get their approval before proceeding.
>
> *It's a commitment.* Although accepting to review is not a legally binding contract, it's an agreement to be taken seriously. Journals depend importantly on assessments in making publication decisions, and ultimately the author's work (i.e. the potential to improve the manuscript) is affected by missing or summary reviews. Reviewers who fail to submit their reports are difficult to replace since this adds significant time (and often not the same expertise) to the review process. Before taking this kind of commitment make sure that your calendar permits. If you accept to review and find that you need an extension, then send an email without delay to the editorial office. A well-run journal will get back to you promptly.

Do as much as you can, but not what you can't. Editors seek scope and depth from a reviewer. Excepting special experts—for example, statisticians—the editor expects each reviewer to assess the full study. With two to four such reports, she can usually make a robust judgment of the submitted manuscript. But the editor realizes that some reviewers will not necessarily have the background to assess all aspects of a manuscript with the same depth. If you agree to review and then find that your expertise is truly limited or lacking on significant aspects of the study, then you should refrain from commenting on those parts, and indicate this in your confidential comments to the editor.

Be consistent between comments to editor and to authors. There is nothing more troublesome for editors than a reviewer who writes upbeat and positive comments for the authors and yet slams the study in the confidential comments to the editor. Author respect *is* a pillar of professional peer review, but this can be problematic for editor and author alike when the report belies the reviewer's frank opinion. This puts editors in the difficult position of justifying what might be a negative decision, despite glowing reviewer comments. Authors evidently do not see a reviewer's confidential comments and are likely to be perplexed (to say the least) by the disconnect between positive reviewer reports and a publication decision to reject.

Consistency can be challenging—a reviewer may find it difficult to be critical in her report to the authors, particularly when young scientists are lead authors or the corresponding author (see also below). Through careful wording, however, a reviewer can be respectful, encouraging and yet honest and accurate; for example: "Despite my criticisms, I believe that the study with additional experiments would make a valuable contribution to the scientific literature."

Look out for young first or corresponding authors. As per above, anything short of a glowing report can be difficult for young authors, and a poorly worded critical report can be devastating. This is not to say that constructive critique should ever be compromised, but rather to express comments carefully and in the most constructive way possible.

Don't make unreasonable demands. Sometimes when reviewing, we find little to quibble about. This can be unsettling, since we may believe that the journal is expecting a major report. It is important not to make demands just for the sake of showing that you have reviewed the paper! Similarly, pointing out limitations is central to reporting, and some of the critiques either will need to be addressed through changes to the text or further analyses or experiments. But reviewers need to take care not to impose requirements (e.g. additional experiments that would be more appropriate as a future study) that go beyond what is reasonable to improve the science. This is particularly important given the tendency of certain journals to accept reviewer reports verbatim and reviewer recommendations on faith, either of which, if unreasonably critical, could result in the rejection of an otherwise acceptable manuscript.

Keep professional and scientific. A final point integrating all of the above is that a reviewer's etiquette must be professional and scientific. Under no circumstances can a report be personal.[5] The line between strongly worded critique and personal references is not always clear-cut. For example, "It is my opinion that this study has important scientific shortcomings" would be a general rebuke of study, but not personal. The same view phrased as ". . . these researchers have obviously failed . . .," or worse ". . . the person who conducted the statistics was clearly incompetent . . ." is personal and unacceptable. Always "review for others as you would have others review for you."[6]

What to Look for in Reviewing a Manuscript

Most journals will be fairly specific about review content in their advice to reviewers. Expect to comment on the following points:

Science. You will evaluate the appropriateness of hypotheses to explain observations, their tests and analysis, and the interpretation of results.

Veracity of facts, logic and claims. A scientific "story" is precise, accurate and verifiable. Facts, logic and claims should be checked or evaluated.

Interest, importance, novelty. What is the advance of the study? Does it change the way we view a problem? How general are the findings?

Clarity. Your job is not to correct grammar and spelling! This said, pristine manuscripts are uncommon. Therefore, you should indeed pay attention to and eventually comment on the overall communicative quality of the study. There may be certain key passages that are unclear and you believe important to include in your comments, but unless you have a lot of time to spare, there is no point in making extensive corrections.

Your Report

Your report to the authors has two main purposes: the identification of scientific limitations and suggestions for improvement/queries. Scientific limitations require some kind of response, a resolution of the problem, a clarification of misunderstanding or a rebuttal when authors disagree with the comment. Improvements and queries are intended to clarify or expand existing material, or consider alternative approaches.

Ultimately, the editor evaluates the importance of a reviewer's comments and—should the journal decide to consider a revision—expectations of how the authors could/should/must revise their manuscript. Reviewers should not expect that each and every comment must be met with a revision. Rather, reviewer opinions or suggestions may result in revisions or, if the authors decide otherwise, simply a justifying reply to the reviewer. In contrast, comments calling facts, claims or methods into question generally require some kind of revision (even if only a clarification), or if the authors disagree, then an explanation or rebuttal. It is the responsibility of the editor to evaluate the adequacy of author replies and eventual revisions.

Here are some basic guidelines in structuring your report:

Start by recapping the manuscript. Write one or a few sentences indicating that you understood the purpose of the study, its main findings and conclusions. The last sentence or sentences in this first paragraph can summarize your overall assessment of the science, its interest and the communicative quality of the manuscript. Some journals explicitly request that reviewers do not include any mention of their publication recommendation in their comments to authors. Others require it. The latter is done for greater transparency in the process and to give more weight to reviewer opinions.

Number your comments, and separate major from minor comments. Dealing with reviewer comments can be a considerable undertaking for authors, involving days, weeks or even months of work. Often, specific comments are clear, as are the revisions expected. But sometimes authors have difficulty telling what comments require a reply, and for those that do, what the reviewer might expect in terms of revisions. It is good practice for reviewers to at least partition what they view as "major" from "minor" comments (and to carefully think of what constitutes "major" or "minor"). This helps distinguish (minor) opinions, observations and suggestions from (major) critiques that require a more consequential reply. Numbering and distinguishing importance also facilitates communication should the reviewer be asked (often months after their original report) to revisit and reassess the manuscript.

You are not *a proof reader.* All too often unwary journals shunt low-standard manuscripts to reviewers. Poorly prepared manuscripts are a waste of a reviewer's time for several reasons: the effort required to understand the science, the temptation to correct or query each instance of incorrect grammar or unclear expression and the real risk that the reviewer misinterprets results. You should contact the editorial office before embarking on your report should you find that you are struggling to understand and fairly assess a manuscript.

Carefully verify revisions. If you have conducted a first round of review and the publication decision was either "revisions" or "reject with resubmission," then it is likely that you will be asked to review the revised manuscript. Reviewing carries this responsibility, since the handling editor is likely not to have *your* expertise, and it could be challenging to accurately verify that the authors have satisfactorily replied to each of your comments with corresponding revisions. In re-rereviewing a manuscript, pay close attention to and verify the adequacy of these elements. If there are issues remaining, then explain these in your report. Sometimes reviewers notice new issues in the study (either ones they previously missed, or others emerging from the revisions). These should be included in your report, preferably mentioning the context (e.g. "In reviewing the revised manuscript, I would like to point out an issue that I previously missed...").

Think carefully before you sign your report. Anonymity is the default of peer review, although the majority of journals will not prevent reviewers from signing their reports. Some reviewers sign their reports as a matter of policy, whereas others never sign reports or selectively sign (usually positive) reports.

Let's examine the important issue of reviewer anonymity in more detail.

Signing a report is more complex than most reviewers imagine.[7] Consider the following scenario. Your report is constructive but critical and indicates in the confidential comments to the editor that the paper should be sent back to the authors for major revision. Nonetheless, based on the ensemble of reviews, the editor decides to reject the manuscript. In receiving a copy of the decision letter, you are probably disappointed. Now imagine the perspective of the authors, who see your report, your name and conclude that *you* are partially responsible for the editor's decision! A level-headed author would ignore this and respect and heed your report in revising her manuscript for another journal. But authors often do not understand how editors arrive at their publication decisions, and, importantly, they . . . are people. Some disgruntled authors will have difficulty dissociating you from their own failure.[8]

My suggestion—especially for younger scientists—is not to submit a signed report unless you are engaged and have thought carefully about the possibilities above.[9]

> *Sign it and then delete it.* One of the many controversies in peer review is reviewer anonymity. Being anonymous promotes frankness, but it can also result in carelessness or being overly critical. This is disrespectful to authors and counterproductive to science. My suggestion is that you write your report and include your name prominently at the top of the document. Revealing your identity will put a check on the severity of your comments and how you actually express them. Then wait a day or two, re-read and eventually edit your report. Make sure the major, valid critiques that would have appeared in an anonymous report are included in your signed report. Take care with your wording to show respect and encouragement. I suggest that you then *remove* your name from the report before actually submitting it. True, although knowing that you will remove your name from the very beginning is not the same as submitting a signed document, it can go some way toward guiding a more conscientious report.

Benefits of Peer Review, Reviewing as a Group and Training

Peer reviewing is not only for the betterment of a manuscript and of science. In being expected to conduct an in-depth assessment, the peer reviewer is obliged to read and think about the manuscript carefully. The peer reviewer comes away informed about a subject, scientific question and its resolution that she otherwise may never read about (but see ethics, below). Moreover, most journals nowadays email the publication decision letter to each reviewer, together with the ensemble of reviewer reports. An interested reviewer can learn a lot by seeing how other reviewers reacted to the same paper, and for journals also providing authors' responses to all reviewers' comments, how authors replied.

Many journals allow group reviews. Reviewing in a group is beneficial both to the editors/authors (in that the review is integration of many complementary views) and to younger members, who will get invaluable experience/training. Before undertaking a group review it is imperative that the invited scientist first secure the agreement of the editorial office, who might ask for the names of all the people participating and that the senior reviewer make explicit to everyone that the review and the manuscript are strictly confidential material.

The usual way we learn peer review is through deduction by examining the reviews of one's own papers. Formal training is uncommon. Beyond the useful approach of co-reviewing with a supervisor or PI, or as part of a group, consider specialized reviewing courses.[10]

Ethics

Reviewing manuscripts carries more responsibility than simply making comments for improvement and grading the work.[11] First and foremost, as a reviewer you are privy to intellectual information. Should you agree to review, the journal will typically either state or refer you to a policy page stating that all information is confidential and that any documents (outside of keeping a copy of your review for your records) must be discarded or destroyed.

Unfortunately, beyond the malevolent use of unpublished ideas and results to nourish one's own research agenda, it is impossible to preclude that a manuscript may have passive influences on reviewers. An honest scientist will do their utmost to adhere to the common-sense ethical notion of respecting the primacy of original ideas and discovery. Respecting primacy means not using ideas for one's own gain, not hurrying one's own work, and not intentionally biasing one's review toward rejection.

Double blind review. The peer review process is open to many positive and negative biases: author identity, home institutes and geographic locations, gender, ethnic background and seniority. With the objective of greater fairness in the peer review process, many journals offer—and sometimes even require—that author names and affiliations be removed from the submitted manuscript. Despite possible advantages[12] according to one study, many authors are not willing to opt for double-blind review, and double-blind papers have less chance of being accepted.[13] Moreover, often, the identity of the authors is either likely or certain based on the article's content, and this may both nullify the intent of double-blinding, and introduce other non-intended biases.

29
Social Media

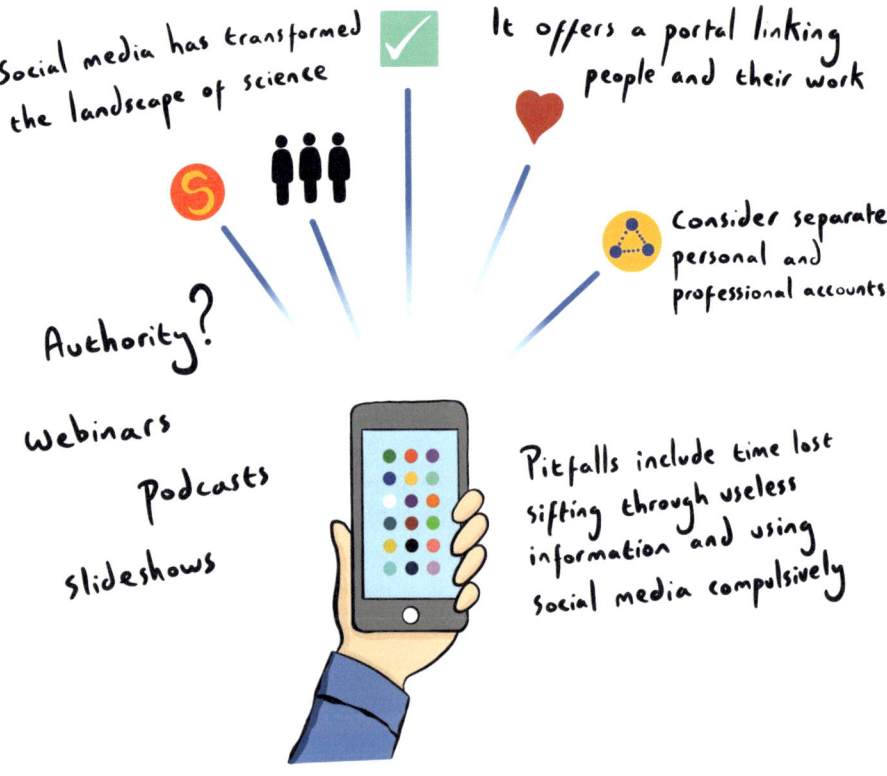

Email, blogs, interviews, podcasts, Facebook, Twitter and conferences are only a few of the many ways scientists communicate with each other, with institutions, journalists and the media, and the public. The ease of use and free access to social media mean that information can spread rapidly and to a massive audience. This chapter discusses how social media and networking can be employed to your advantage and some possible traps to avoid.

Think for just a minute from where we have come. In 1990—only about 30 years ago—email was just emerging. For most scientists, communication was principally in person, through letters and by telephone. You would never send a tweet-like message via snail mail, nor call someone on the phone every time you read an interesting paper. In 1990 and indeed until the Internet explosion, communication took more time and effort—we had to think harder about what was useful and important.

Things have clearly changed.

Social media allows virtually instant access to most any kind of information of interest (and not of interest), and enables you to share news, that of others, give your opinion or engage in active dialogue in a matter of seconds or minutes.[1] Not only is social media easily accessible and user-friendly, but most of what you can—or would ever want to—do can be done with a handful of tools and interfaces: the Internet, cell phones, personal websites, blogs, social networks, emails and text messages.

Social media can also be abused. With a simple spoken phrase or drag of the cursor you can ricochet back and forth between actual work and the latest research news. Concentration is easily broken each time you have the urge to see what others are talking about on Twitter or just feel like checking whether there's any new email in your inbox.

Being connected opens new doors, but also poses formidable challenges.

Social Media Puts You in Touch

Social media is the universe of dedicated servers and personal blogs that permit channels of communication—or rather—the flow of information. Social media continues to diversify and includes social networking, social bookmarking, blogging, microblogging, wikis, and media and data sharing.[2]

Until the Internet age, science was mostly communicated through the reading and citation of published articles and attending and giving research seminars. Tens or hundreds of people may have read a published article or attended a talk, but most of them were probably not venturing very far from their own areas of research.

The Internet and social media have radically changed this. Rather than being limited to a bounded group, the number of *potentially* connected people on the Internet is virtually limitless. This is not to say that 1 million people will ever be interested in any of my publications! However, this number could conceivably be 1000 or more, whereas without the Internet it would have been 10 or perhaps 100.

Consider the following illustration.

I have published an exciting paper and hope that it will be noticed in the scientific community. The usual channels for this are database searches, citation in other publications, or when I present the paper at conferences or invited seminars. With social media—take for example Twitter—all I have to do is tweet a message announcing the paper. The more followers I have, the more my message will be seen and the more people will actually look at my paper. But because some of my followers believe it is important to "spread the word," they, in turn, retweet my message—possibly with their own recommendation in the form of a brief comment—to *their* followers. And so on...

Large numbers of followers and the chance effect of one retweeter being a "super-spreader"—that is, having thousands or more followers—can mean that retweets of my original message grow exponentially. In a matter of minutes, hundreds or thousands of people may see my article. This is *not* to say that many of them are interested scientists, but nevertheless some will be.[3]

Some Pros, Some Cons

Social media interfaces provide a world of possibilities.[4] Again, take Twitter. Within minutes a scientist can do one or more of: tweet one's own messages, retweet the messages of others, enter into a discussion thread and obtain information from a range of sources, including individuals, academic departments, research teams, journals, etc. Social media such as Facebook and Twitter are particularly useful in engaging interactions among scientists, stakeholders and the public.[5] More sophisticated and complete webinars and podcasts are ways to share your own work, and learn about that of others.[6]

Social collaboration networks connect academics with common interests. These range from professional networks such as LinkedIn, which advertises your CV, accomplishments and has a live news feed, to more scholar-tailored networks such as ResearchGate in the basic sciences, Academia.edu in the social sciences and humanities, and most recently the non-profit ScholarlyHub. These platforms allow the posting of papers and promote connections.[7]

But the mind-boggling variety of possibilities in social media highlights what are perhaps its four main drawbacks and limitations: information overflow, user addiction, and questionable authority and impact.

> *Information overflow.* Knowing which sources to follow, what posts to read and which to give weight to is as much an art as a science. If you are on Twitter, and follow 100 people, journals and news media, then you are probably receiving hundreds of tweets each day. Beyond the problem of scrolling through this information and deciding what to read, there is the issue of finding *too much* of interest. Similar to doing a literature review, a huge number of social media entries may appear to be essential reads. Learning how to be selective is part of how one more generally deals with information overflow.
>
> *User addiction.* Arguably, the main challenge for those who use social media is to know when to start and when to stop. Some actively seek media as a pleasurable and useful outlet for learning about and communicating research. Others use it more passively as an easy, quick and entertaining refuge from work. The use of social media—be it on a telephone or screen—is largely a private affair between you and your device. This means that only *you* can stop media from becoming a time sink.
>
> *Authority.* Social media potentially puts scientists into touch with orders-of-magnitude more people than they've ever actually met and sources than they would

have otherwise consulted. This presents the serious issue of receiving what may appear to be factual information, but really can be anything from intelligent guesses to the persuasive phrasing of total nonsense. Unfortunately, the interest, importance and authority of a source cannot be assessed simply based on the number of *likes* or *retweets* it receives, since many viewers mechanically approve what others already have, creating a snowballing effect—the tweet goes "viral." Accurate filtering for authority is next-to-impossible, since even "trusted" sources may relay dubious information. The more vigilant learn how to spot this, whereas the less vigilant risk not only giving credence to unmeritorious information, but also spreading it to others.

Impact. Although certain measures of article influence such as PageRank and the Eigenfactor are correlated with journal impact[8], there is mixed evidence that employing social media is similarly associated at the article level.[9]

Blogs

Blogs developed as extensions of websites. As users realized the opportunities provided by blogging, dedicated sites grew exponentially. Now anyone can easily create her own blog via a developer. Blogs provide a forum for talking about most anything that pleases the blogger, such as particular scientific themes or the interface between science and society or policy. Blogs can be written for anything from specialized audiences to the general public.

Regular blogs may garner a dedicated following of hundreds or more. They can therefore serve to promote one's own research (or research agenda) in a more user-friendly way than technical papers. Blogs need not be dedicated websites: many scientists use other social media such as Twitter and Facebook as a microblogging platform. Some blogs are one-way, that is the blogger publishes essays with no possibility for comments. Others let viewers freely comment, and still others require that respondents register so as to have filtering capabilities. Blogs with user comments are particularly useful for scientists to get feedback on periodic installments that may become a future publication.

Still New

Social media is part of our everyday lives, yet we often forget just how young most platforms actually are. Two of the most commonly used (Facebook and Twitter) are barely 15 years old.

Social media is used for a variety of reasons (or sometimes no particular reason), is incredibly diverse and continues to diversify.[10] Because there is little or no filtering of any kind—no oversight and sometimes not enough self-restraint—social media is still in many ways the "Wild West." Perfectly unheard-of people have huge numbers of followers. Anyone can come along and add an unpleasant comment to a useful post.

Misleading or aggressive information is common and can be erased at the push of a button. Self-aggrandizing is commonplace.

All of the above contributes to an atmosphere that is the antithesis of neutrality and scholarship, so prized in scientific publication.

My recommendation is to treat social media like other forms of communication. You are a professional and keep to the science at meetings and conferences—I suggest that you approach social media no differently.

Finally, social media—and more generally information networks—goes well beyond the scientist-centered world that we all know. Institutions, research administrators and librarians, funders and governments are all keenly interested in the value of research and in providing indicators to stakeholders and the general public. Accomplishing this requires the collection, processing and packaging of monumental quantities of diverse data. This is a burgeoning part of Open Science that promises to change the evaluation culture. Stay tuned.

30
Old Dogs, New Tricks

The world of scientific communication is changing fast and in unpredictable ways. This goes from the media we use, through our motivations in publishing, to how we can make science a more collective and "scientific" enterprise. This final chapter reviews the book's main messages, current issues in scientific communication, and underscores the importance of educating young researchers in writing and publishing science.

Communicating science is both a world unto itself and part of something larger—a community of scientists—but also collaborators, journals, publishers, academic societies, institutions, institutes, librarians, governments and the public. How these actors and stakeholders interact will be key to the future of science, be it what gets funded and where and under what conditions it is published, who has access to what information, and who gets the jobs and career rewards. This book is an introduction to this complex and rapidly changing world.

In closing, I would like to underscore the book's main messages and present views on the future of science communication.

Use *Models*

Scientific writing is technical, follows a number of norms, but is readily mastered with method and practice. Speaking for myself, I started feeling comfortable writing papers at the end of my PhD. I was a very avid reader, and noticed that some papers "spoke to me" more than others in the way they were written. The logic, the clarity and the story. I carefully read a select few to try to understand what made them work. I saw particular features that turn out to be shared by all *Models*. Paragraphs begin with an introductory sentence, followed by a logical series of sentences developing the subject. Every paragraph has a concluding statement. There are clear patterns in how each section is written, how previous work is cited, how the choice of title reflects the overall mood of the story and the key punctuating role of that first paragraph of the Discussion. Sentences were clear and concise, and flowed seamlessly from one to the next. Carefully reading perhaps five or six such *Models* gave me the necessary elements to write my own papers. I have used *Models* ever since, and indeed used them to inspire certain passages of this book.

Models extend beyond published papers. Knowingly or not, you are potentially influenced by speakers at conferences, students or postdocs at your institute, more senior scientists, mentors and PIs. Most interactions will change little or nothing, but occasionally someone will have a major impact. It could be the way a scientific question is formulated, a seminar is presented, a journal for submitting a paper is chosen or a grant proposal is structured. We underestimate the importance of social learning in science. Being attentive to or even searching for inspiration is—like the use of *Models* in writing—a key to success, and the basis for the last message below. Just like writing *Models* in Chapter 7, think carefully about why an approach works before actually adopting it.

Journal Choice Does Matter ... at Least for the Time Being

There are those who say "It does not matter all that much where you publish ... interested readers will find your work." This has some truth, but is not the whole story. Yes, good science stands on its own "legs," but different people find published articles in different ways. The science that pops up on a Google search is not always the same as that seen on Twitter, and certainly both are far less curated than the Web of Science or Scopus. Perhaps more troubling is that your science will elicit different responses—at least in the eyes of some—depending on your name, in what journal your paper is published and how many times your study has already been cited. Although we can educate researchers on the pitfalls of judging science by proxies such as journals, the reality is that scientists are largely free to decide what they read and cite. As related in Chapter 23, evaluation committees too use names—and more specifically metrics that they believe reflect the importance of journals, articles and scientists themselves. Thus, if your objectives include impact and career, and impact and career continue to depend on proxies such as journal prestige and impact factor, then where you publish *does* matter.

Journal choice is important for young career scientists in today's world, but is likely to be less so in the future. Pledges[1] to make science more open, and base evaluations on scientists and their science and less on journals are being followed closely by stakeholders[2] including funding agencies, governments, publishers, the public and the scientific community. Nevertheless, journal names are unlikely to disappear from evaluations any time soon. Journals have legacies—in particular, readership, reputation and prestige. Successful articles (and their authors) effectively "inherit" the journal's legacy. As long as journals differ, so too will the perception of the papers they publish. Evaluation committees currently have no method for removing the influence of "journal" from measures of impact. The best they can do is understand the limitations and biases in metrics and the source data, and decide how this affects the weight given to these numbers in their overall evaluations.

The Open World, Slowly …

Open Access (OA)—the flagship of Open Science—is intuitive and appealing because it fosters communication and innovation. OA is moving forward, but more slowly than many expected. As I write, about 30 percent of published articles are OA,[3] and although growing, it is not known what level it will attain or even if the APC economic model is viable in the long term,[4] or the extent to which future OA journals will be Gold, Green or hybrid.

Why is the move proving difficult? Despite widespread approval, two of the main actors in the shift to OA—authors and publishers—are sometimes hesitant or even flatly opposed. Authors want to choose where they publish, and as discussed in Chapter 16, many make this choice based on journal readership, reputation and prestige—they only consider OA secondarily. Important here, and more generally, is whether the author can, or is even willing, to cover the APC.

Although publishers usually adopt full OA for new journal launches, they are largely opposed to flipping established subscription-based titles for the simple reason that they would lose considerable revenue. As discussed in Chapter 22, established journals (which also tend to be those with the highest reputation or prestige, and therefore the largest subscription bases) would need to charge high if not exorbitant APCs to equal the revenues they receive from subscriptions. Moreover, the demand for subscriptions is indirectly facilitated by both the evaluation culture and scientists themselves: as long as scientists publish important work in prestigious subscription-based journals, there will be demand to read these articles and for librarians to subscribe to the associated journals. Many established titles have nevertheless shifted partly to OA—Green, Delayed or hybrid. Hybrid generates considerable additional revenue (and tends to be more expensive than Gold OA), whereas Green and Delayed with their constraints provide a second-class alternative to Gold. Neither Green nor hybrid journals can become 100 percent immediate OA, since subscriptions would be pointless.[5] This means that even with growth in the numbers of new Gold OA journals—barring a shift in attitudes (but see Plan S below)—OA will never attain 100 percent.

Several initiatives aim to make science more open, and more research and researcher-centric.[6] OA2020 and DORA in particular see a healthier world for science and for scientists as a collective, once products are accessible and credited based on content and not where published. Many stakeholders support OA,[7] including the Wellcome Trust, the National Institutes of Health (NIH), NSF International and the Bill and Melinda Gates Foundation, the US National Academies of Sciences, Engineering, and Medicine, UK Research and Innovation, the European Commission, the United Nations Educational, Scientific and Cultural Organization (UNESCO), the Organisation for Economic Co-operation and Development (OECD), major initiatives notably in Latin America, and a long list of research institutes worldwide. Some have gone as far as making pro-OA policy changes.[8] More recently, TSPOA (Transitioning Society Publications to Open Access) has started an initiative that "…will serve as a communication forum, clearinghouse, and advisory group in providing support for libraries, publishers, and scholarly societies working toward transitioning scholarly publishing from subscription-based to open access."[9]

The boldest initiative to date is Plan S—an international coalition of institutes, commissions, funding agencies and learned societies that will apply the central objective of immediate OA publication starting 1 January 2020, and under certain conditions immediate OA in hybrid journals until a review in 2023. Plan S specifies acceptable journal models, and is *de facto* decreed and enforced by members of the coalition to their constituent researchers. Given that approximately 85 percent of journals would be off-limits,[10] including prestigious titles such as *Science* and *Nature*, many scientists and institutes have expressed concern.[11] We don't yet know what the impact of Plan S will be on the subscription model, or to what extent researchers concerned will adhere, and if complying, decide to publish the definitive versions of their articles as Gold OA or, as explained in Chapter 15, on free-of-charge preprint platforms.

Meta-research

Perhaps the greatest irony in science is that some of it is demonstrably substandard: sloppy, biased and occasionally fraudulent. Substandard research is able to get published and subsequently go unnoticed, because our mechanisms for reliably identifying it and calling it out (correcting, rejecting or commenting) are limited. Various initiatives such as collaborative and post-publication peer review are beginning to change this, but we still have a long way to go in developing and applying the tools to scientifically examine science.

Meta-research is the scientific study of scientific research itself with the objective of evidence-based improvement.[12] It ranges from how studies are conducted, analyzed and reported, to the usefulness of analytics for evaluating research impact, to uncovering biases, such as p-hacking, lack of reproducibility and self-citations. Meta-research naturally meshes with Open Science, since the former relies on products of the latter: information access, use and transparency. The two are also related since both view science as open to constructive critique and improvement. Meta-research and Open Science together promise a sea change in the way we conduct, write and publish science.

Pass it on

This book is based on my own experiences as a practicing scientist and chief editor. As stated in the Preface, there is no single, best method or tested advice to writing and publishing science. This book is intended to serve as a basis to develop your own writing methods, and your views on publishing and the future of how we communicate science. Discussion with mentors, PI and colleagues is the surest way to improve your skills, and to prepare yourself for teaching this material one day to your own students.

That writing science is so challenging for many and the world of publication can be a black box, owes to the fact that the art of scientific communication is rarely taught in formal courses, seminars and discussion groups. It is my hope that this book serves as both a personal guide and reference, and a teaching support for instructors to illuminate and engage their students in the world of scientific writing and publication. Integral to this project is teaching how Open Science is enacting major changes, and the importance of being responsible scientific citizens.

Students—more than ever—need guidance for communicating science.

Glossary

Abstract. Synopsis of the paper, summarizing each major section.

Academic society. An organization that promotes one or more academic disciplines, sometimes including overseeing an academic journal. Also called "learned societies."

Acceptance. The formal notification that a submitted manuscript has been accepted for publication in a journal.

Acknowledgments. The section or passage in a paper, where individuals and organizations are recognized for having contributed to the execution, funding or comments and discussions leading to the published paper.

Altmetrics. An alternative to citation metrics, sometimes referred to as "influence," that scores multiple forms of online research output.

Analytics and big data. The analysis of large and complex data sets relating to disciplines, journals, articles or individuals that yield insights into characteristics and performance.

Appeal. The request by authors of a rejected manuscript for the journal editors to reconsider the decision.

Arbitration. The act of an editor to scrutinize divergent reviews so as to come to a justified and coherent publication recommendation for the chief editor.

Archiving. The deposit of data, usually used to achieve a published study, for future use onto a dedicated electronic server.

Article Processing Charge (APC). The charges levied by journals on authors upon manuscript acceptance, for publication as immediate Open Access. APCs are charged to the authors, but are typically covered by the author's research funding or institution.

Article. The published version of a study in a peer reviewed, scientific journal. Also referred to in this book at the "publication."

Assessment. The in-depth analysis by an expert: generally used in this book to mean either by a reviewer or the editor on a submitted manuscript.

Associate editor. A member of the editorial board who has more responsibility (e.g. overseeing a particular section of the journal) than general members. Also called a "senior editor."

Author. An individual contributing to and taking responsibility for a study, so as to merit inclusion on the credited list.

Bias. Intentional or unintentional favor resulting in unfair credit or influence. Bias in science can involve, for example, gender, institution or country, career stage or perceived status.

Big deals. The massive journal packages proposed by large, for-profit publishers starting in the 1990s, to be a cost-effective way for libraries to replace multiple, single-title subscriptions.

Capping. The establishment of a maximum article processing charge (APC) for immediate Open Access journals.

Cascade journal. A journal that receives submissions that have been previously rejected by another more selective title, from the same publisher.

Chief editor. The person responsible for a journal's publication decisions. Also referred to as the editor in chief.

Citation metrics. Citation-based performance characteristics of a subject, be it a researcher, article, journal, discipline, etc. The impact factor is the most widely known.

Citation. The explicit reference within an article to a previous or concurrent publication. The detailed reference corresponding to the citation is provided, usually on a list at the end of the article in a section called "References" or "Literature Cited."

Collaboration. Two or more individuals or research groups that combine skills to conduct or report research.

Confidentiality. The principle whereby editorial office staff, editors and reviewers are not to discuss, outside of the journal, any information relating to submitted manuscripts.

Conflict of interest (COI). Financial, professional or personal interests that can influence or bias research or the assessment of research.

Copyright. The individual(s) or organization(s) who holds the license of legal rights to a publication as intellectual property or for reuse.

Corresponding author. The author to whom all correspondence is addressed during the submission process. This information is provided during the online submission and in the cover letter. Depending on the journal, the published version of the article may declare one or more corresponding authors (with their email addresses), whom interested readers can contact.

Cover letter. A document that describes the study, its significance and any requests such as preferred or barred reviews. A cover letter also accompanies any manuscript revision, recounting the publication decision and the main points for the editor's attention.

Creative commons (CC). A non-profit organization that provides free licenses that specify the rights reserved to creators and those waived for reuse. The CC license is the owner's modification of their existing copyright terms. In CC0, copyright waved and data are in the public domain. CC-BY has the added requirement that users must credit the source.

Data sharing. The release and free access of research data for use, reuse or redistribution, with the main condition being explicit attribution (either names of creators or citation of source publication). Also referred to as "open data."

Decision time. The time elapsed between acknowledgment of manuscript submission and the publication decision.

Delayed Open Access journal (Delayed OA). Open Access made available by the publisher on the journal's website after an embargo period. Sometimes confused with "Green OA."

Desk rejection. A journal's decision to reject a submitted manuscript without peer review.

Digital Object Identifier (DOI). A unique identifier for objects such as data sets and journal articles. DOIs facilitate identification, access and association with information in the form of metadata.

Directory of Open Access Journals (DOAJ). A community-curated directory to peer reviewed, high-quality, Open Access journals.

Discussion. The last section of an IMRaD (Introduction, Methods, Results and Discussion) paper, which presents the relevance of findings to current knowledge and understanding, and may also set out future directions.

Double-blind review. Peer review in which both the authors' and reviewers' identitics are concealed.

Editor. Any member of the editorial board, including the chief editor and associate editors. Usually used in this book to refer to the handling editor.

Editorial board. A college of dedicated experts representing the thematic scope of the journal. Editors endorse the journal, handle manuscript submissions, advise the chief editor on policy and thematic directions, and approach potential contributors to submit their finest work.

Editorial office. The nucleus of a journal that ensures the execution of its workflow.

Embargo period. The minimum time required after publication for an article from a Green OA or Delayed OA journal to become available to non-subscribers.

Evaluation culture. Overemphasis on the importance and use of journal names, ranks and citation metrics in the evaluation of science and scientists. Also called the "prestige culture."

Gold Open Access journal. Immediate OA on journal website. The author retains copyright. APCs apply in most reputable journals.

Green Open Access journal. Subscription-based journals that allow the posting of the article as OA after an embargo period on a repository. The copyright is usually assigned to the publisher.

Handling editor. A member of the editorial board who is assigned the responsibility for the assessment of a manuscript (based on peer reviews) and publication recommendation to the chief editor.

Hybrid journal. A subscription journal that provides an option to publish immediate OA with an APC.

Immediate Open Access. The immediate posting of an OA article for publication on the journal website. Includes Gold OA and OA published in hybrid journals.

Impact. The influence of a journal, article or researcher on science as measured by citations.

IMRaD. The organization of most original research papers into separate Introduction, Methods, Results and Discussion sections.

Influence. The influence of a journal, article or researcher on science. Depending on the metric can be measured using citations, journal rank or online access statistics to websites, downloads and social media.

Introduction. The first section of an IMRaD paper. Provides the context of the study and specific problem or question addressed.

Jargon. Terminology that is only understandable by a restricted set of readers.

Journal. The regular publication of a collection of peer reviewed papers that have passed the standards ensured by an editorial board. Also referred to as "title."

Journal Impact Factor (JIF). A specific calculation of impact for journals covered by the Web of Science, based on the average number of citations per article. More generically called the impact factor (IF).

Journal rank. The ranking of a journal in its discipline according to its impact factor.

Managing editor. The person responsible for the running of the editorial office and the journal workflow.

Manuscript. A scientific paper after submission to a journal, but before publication. In this book, I also sometimes refer to a paper prior to submission as a "manuscript."

Manuscript handling system. The online interface that ensures all automated steps of manuscript handling, from submission to publication. Such systems are tailored to the specifics of journal workflow, maintain reviewer databases, archive all manuscripts and associated correspondence, and can perform Internet searches.

Media embargo. A requirement at certain journals (e.g. *Nature*, *Science*) not to discuss the content of an accepted article with the media until a specified time prior to publication.

Megajournal. An Open Access journal that accepts manuscripts based solely on scientific and technical soundness.

Metadata. Electronic information such as the title, author names, abstract of an article that makes it discoverable and its contents eventually usable for data sharing.

Methods. The section in IMRaD where sufficient details are provided about experiments and analysis so that the reader can replicate the study. Also called "Materials and Methods."

Minimum publishable unit (MPU). The minimum amount of information in a manuscript that passes publication standards at a given journal.

Open Access (OA). The free, unrestricted access to online research outputs, often with rights to reuse.

Open Peer Review. Peer review in which the reviewer identities are disclosed to the authors. Depending on the journal, the reviews may or may not be published online with the article.

Open Science. The umbrella of the open movement, including Open Access, Open Peer Review, data sharing, etc. Open Science seeks to make as much science as possible freely available, usable and reusable.

ORCID iD (Open researcher and contributor ID). The unique, persistent identifier associated with a unique individual author.

Original research article. A publication based on data collection and analysis by its authors. Also called a "primary research article."

p-hacking. Biased or fraudulent analysis of datasets that misleadingly finds or emphasizes statistically significant effects.

Paper. Used as a generic term for a preprint, manuscript or article.

Paywall. Access to a publication that either requires a subscription or one-off payment.

Peer review. Review of a submitted manuscript by independent experts.

Post-publication review. A form of Open Peer Review in which reviews are posted online, usually associated with preprints, but also version of record articles via dedicated platforms.

Predatory journal. A business model whereby a group posing as a legitimate journal receives and publishes manuscripts from sometimes unsuspecting authors, charging what is usually an unusually low APC. Also called "predatory Open Access journal."

Preprint. A version of a paper that is made immediately and freely available on the Internet via a dedicated platform.

Prestige. High esteem by the scientific community, as reflected by journal ranking or impact factor.

Principal investigator (PI). The lead researcher of a funded project or head of a research group.

Priority. The earliest publication of discovery, officially recognized through citation. Also called "precedence."

Production. The publisher's department responsible for transforming an accepted manuscript into the final article, including typesetting, author corrections, online publication and indexing.

Profit. The net earnings of a for-profit publisher after all costs are deducted from total revenue. See also "surplus."

Proofs. The stage in the publication process where the manuscript has been typeset and the authors make final checks regarding the accurate and clear reproduction of their accepted paper.

Protocol. The decision-making process used by a journal to advance a manuscript through each phase and contingency in the workflow.

Publication decision. The decision made by the chief editor to either accept, reject or revise a submitted manuscript.

Publisher. Ensures some or all of the non-scientific aspects of journal function, including financing the journal, overseeing the editorial office and production, managing the website and publicity, and hosting pdfs and metadata.

Quality. The level of excellence compared with a standard.

Readership. The segment of the scientific community that preferentially browses its contents.

Recommendation. The handling editor's opinion sent to the chief editor regarding whether and why a manuscript should be accepted, rejected or revised.

References. The listing of all citations typically at the end of the article. Also called "literature cited" or "bibliography."

Rejection without prejudice. A soft rejection in which the editor specifies that the decision to reject will not influence the consideration of any resubmission.

Rejection. The publication decision not to publish a submitted manuscript.

Reproducibility. Finding the same result when repeating an experiment using the same methodology.

Reputation. The shared view in the scientific community that a given journal has or lacks certain standards.

Resubmit. The willingness of the journal either to give further consideration to your manuscript (revise and resubmit) or to consider a new version (reject and resubmit).

Results. The section in IMRaD that presents the findings of the study.

Reviewer. An expert who contributes an assessment or "review" of a submitted manuscript. Reviewers for most journals are independent of the editorial board. Also called a "referee" or "peer reviewer."

Revision. The decision to consider a revised and resubmitted version of the manuscript in response to peer reviews and editorial comments.

Salami slicing. The division of a single study into more than one publication, which may or may not be MPUs.

Science citation index. The list of journals covered by the Web of Science platform.

Scientometrics. The study of the quantifiable characteristics of scientific research.

Senior author. Typically, the PI or a senior scientist who organized and oversaw the study. This person is usually listed last on the author list.

Serials crisis. Annual journal subscription cost increases that are beyond the ability of many libraries to maintain all of their titles.

Soft rejection. The decision to reject a paper, but with willingness or encouragement to consider a resubmitted version.

Stakeholder. Individuals and organizations concerned with scientific research. This includes funders, research institutes, publishers, governments and the public.

Standard. The required level of a quality criterion for editorial acceptance.

Submission. The state of a manuscript while it is being handled by the journal.

Subscription. The right to unlimited access to journal content, based on a fee paid by a consortium, institute or individual.

Surplus. The net earnings of a not-for-profit publisher after all costs are deducted from total revenue. See also "profit."

Terminology. The use of specific words to facilitate communication.

Title. The most succinct description of the study that headlines an article. Also used in this book to refer to a journal.

Version of record. The final, official published version of an article.

Waiver. Either a reduction or cancellation of subscription or publication costs, issued under certain conditions by a publisher.

Web of Science (WoS). An online subscription citation indexing and analytics service owned by Clarivate Analytics.

Workflow. The method for handling a manuscript from submission to the publication decision, tracked by the manuscript handling system and overseen by the editorial office.

Notes and References*

Chapter 1

1. Royal Society of London. Order in Council. March 5, 1665.
2. Vale, R.D. & Hyman, A.A., 2016. What defines "priority of discovery" in the life sciences? *ASAPbio*, http://asapbio.org/priority

Chapter 2

1. Impact is notoriously hard to define and to represent as an unbiased metric. I use impact in this book to mean how a published article changes science. The measurable unit is the citation, which is in turn the basis for a variety of impact metrics. These points are discussed in Chapter 23 ; see also Sugimoto, C.R. & Larivière, V., 2018. *Measuring Research: What Everyone Needs to Know*, Oxford University Press, Oxford.
2. Fidler, F., Chee, Y.E., Wintle, B.C., et al., 2017. Metaresearch for evaluating reproducibility in ecology and evolution. *Bioscience*, 67(3), pp.282–289.
3. A highly productive group can even adopt a culture of questionable (but sufficient) quality, if productivity is key to success; Smaldino, P.E. & McElreath, R., 2016. The natural selection of bad science. *Royal Society Open Science*, 3(9), p.160384.
4. Vale, R.D., 2015. Accelerating scientific publication in biology. *Proceedings of the National Academy of Sciences of the United States of America*, 112(44), pp.13439–13446.
5. https://sfdora.org/

Chapter 3

1. Tenopir, C., Volentine, R. & King, D.W., 2012. Scholarly reading and the value of academic library collections: results of a study in six UK universities. *Insights: The UKSG Journal*, 25(2), pp.130–149.
2. Bornmann, L. & Daniel, H., 2008. What do citation counts measure? A review of studies on citing behavior. *Journal of Documentation*, 64(1), pp.45–80.
3. Greenberg, S.A,, 2009. How citation distortions create unfounded authority: analysis of a citation network. *BMJ*, 339, p.b2680.
4. Himmelstein, D.S., Romero, A.R., Levernier, J.G., et al., 2018. Sci-Hub provides access to nearly all scholarly literature. *eLife*, 7, p.e32822. See also https://unpaywall.org/ https://kopernio.com/ https://openaccessbutton.org/
5. https://www.semanticscholar.org/
6. https://i4oc.org/
7. Clearly, about 90 percent of the articles had one term or the other, but not both. Herbivory is the general phenomenon of eating plant material, whereas defoliation refers to the complete

---

* All URLs accessible on March 1, 2019.

consumption of leaves by herbivores. While defoliation implies herbivory, the reverse is not necessarily true.
8. Other databases include Scopus, Medline and Google Scholar.

Chapter 4

1. "Suspicious" could mean a single long string (such as one or more complete sentences), or multiple phrases each of several or more words. Many journals set a threshold above which they consider content to be potentially plagiarized. To avoid plagiarism, authors should place any copied text into quotations and cite the original source, even if it is their own previous publication.
2. It is only in the past 10 years that journals routinely check for plagiarism. Butler, D., 2010. Journals step up plagiarism policing. *Nature*, 466(7303), p.167.
3. See Association of Learned and Professional Society Publishers, www.alpsp.org/default.htm
4. This is loosely based on an incident that occurred at *Ecology Letters* in 2005. See: ERRATUM. (2007). *Ecology Letters*, 10, 435–435.
5. https://publicationethics.org/

Chapter 6

1. This is not completely true. Before actually writing, you will have outlined your paper, conducted a literature review, written at least a draft of your Methods and (nearly) completed your analyses for the Results.
2. There is some evidence that narrative style correlates with impact, as measured by citation frequency. Hillier, A., Kelly, R.P. & Klinger, T., 2016. Narrative style influences citation frequency in climate change science. *PloS ONE*, 11(12), p.e0167983.
3. Numerous collaborative writing and document sharing tools exist, for example GoogleDocs and (for LaTeX) Overleaf.

Chapter 7

1. *Proceedings of the Royal Society of London* B 282, 20152207; numbered references are removed and American English used.

Chapter 8

1. Points 2–4 are particularly important and treated in detail in the next chapter.

Chapter 9

1. Bornmann, L. & Mutz, R., 2015. Growth rates of modern science: A bibliometric analysis based on the number of publications and cited references. *Journal of the Association for Information Science and Technology*, 66(11), pp.2215–2222.
2. Be careful with the use of color. For those relying on black and white printers, the main messages in your figure may be lost. Consider using a color gradient with corresponding labels so that even if the figure is printed in black and white, the reader can navigate. Also try to make

your figures color-blind friendly. Cox, L. 2015. Tips for designing scientific figures for color blind readers. Accessed at: www.somersault1824.com/tips-for-designing-scientific-Figures-for-color-blind-readers/.

Chapter 10

1. And, of course, you should cite this work!
2. If your study is structured to answer a specific question or test a particular hypothesis, then *a posteriori* results must not modify this.

Chapter 11

1. Elements can also be *deemphasized*, such as when sample sizes are small (and the result preliminary), or a result is straightforward or obvious.

Chapter 12

1. Mabe, M.A., 2009. Scholarly publishing. *European Review*, 17(01), p.3 (and references therein).
2. There are some attempts to make journals more transparent, at least with regard to peer review, e.g. www.peere.org/peeer-in-a-nutshell/; http://www.bbk.ac.uk/news/birkbeck-to-investigate-the-peer-review-process-in-new-research-project
3. Digital Object Identifiers (DOIs) give a unique tag to an article, report or dataset. Before deciding on a journal to submit your work, verify that the publisher applies DOIs to their published articles. Most large publishers are members of Crossref, a registry organization for DOIs.
4. Helmer, M., Schottdorf, M., Neef, A., et al., 2017. Gender bias in scholarly peer review. *eLife*, 6, p.e21718.
5. Many disciplinary journals will facilitate the process by guiding authors in their revisions.
6. Wakeling, S., Spezi, V., Fry, J., et al., 2017. Open Access megajournals: The publisher perspective (Part 1: Motivations). *Learned publishing: journal of the Association of Learned and Professional Society Publishers*, 30(4), pp.301–311.
7. Björk, B.-C., 2015. Have the "mega-journals" reached the limits to growth? *PeerJ*, 3, p.e981; economies of scale also applies to large versus small publishers, which is why many society publishers partner with big publishers.
8. Fang, F.C. & Casadevall, A., 2011. Retracted science and the retraction index. *Infection and Immunity* 79(10), pp.3855–3859. Tressoldi, P.E., Giofré, D., Sella, F., et al., 2013. High impact – high statistical standards? Not necessarily so. *PloS ONE*, 8(2), p.e56180. Macleod, M.R., McLean, A.L., Kyriakopoulou, A., et al., 2015. Risk of bias in reports of in vivo research: A focus for improvement. *PLoS Biology*, 13(10), p.e1002273. Bohannon, J., 2013. Who's afraid of peer review? *Science*, 342(6154), pp.60–65.
9. Under my tenure at *Ecology Letters* we always marshaled three reviews, and invited a fourth reviewer if there was any concern regarding limited reviewer expertise.
10. Assessor scores are at most weakly correlated, and weakly predict the eventual number of paper citations. Eyre-Walker A. & Stoletzki, N., 2013. The assessment of science: The relative merits of post-publication review, the impact factor, and the number of citations. *PLoS Biology*, 11(10), p.e1001675.
11. Murray, D., Siler, K., Larivière, V., et al., 2018. Gender and international diversity improves equity in peer review. *BioRxiv* DOI:10.1101/400515

Chapter 13

1. Eyre-Walker & Stoletzki, 2013, *op. cit.*
2. This would typically be accompanied by the willingness to reconsider the manuscript if revised to meet journal writing standards.
3. https://peercommunityin.org/; https://www.authorea.com/inst/14743-prereview

Chapter 14

1. If you do not receive email confirmation that your manuscript has been submitted, I suggest waiting a few days before contacting the editorial office. Note that many manuscript handling systems now indicate "out for peer review" online, rather than notifying by email.
2. Many journals have reduced or no copyediting and apply house styles. For non-English speakers, journals may have a list of external services that can help improve the English. Authors should make sure that any editing services they themselves seek are reputable, and should let the editorial office know if they had help.
3. Setting precedence can be affected by the editorial process itself (delays, biased reviewers), and is of particular concern for papers in the molecular and biomedical sciences. Preprints have an important role to play here, since they enable authors to claim precedence and get feedback on their paper (see Chapter 15).

Chapter 15

1. Cited in: Johnson R., Watkinson A. & Mabe M., 2018. *The STM Report. An Overview of Scientific and Scholarly Publishing.* 1968–2018. Association of Scientific, Technical and Medical Publishers, The Hague.
2. National Science Board, 2018. *Science and Engineering Indicators* 2018. NSB-2018-1. National Science Foundation, Alexandria, VA. Accessed at: https://www.nsf.gov/statistics/indicators/
3. www.budapestopenaccessinitiative.org/read; https://openaccess.mpg.de/Berlin-Declaration; Guédon, J.-P., 2017. Open Access: Toward the Internet of the mind. Accessed at: www.budapestopenaccessinitiative.org/open-access-toward-the-internet-of-the-mind
4. There are several broadly similar conceptual models for Open Access. See, for example, the DART framework (Discoverability, Accessibility, Reusability and Transparency). Anderson, R., Denbo, S., Graves, D., et al., 2016. Report from the "What Is Open?" Workgroup. *Open Scholarship Initiative Proceedings*, 1, pp.1–5. Accessed at: http://dx.doi.org/10.13021/g8vg6g
5. Johnson et al., 2018, *op. cit.*
6. Delayed OA (or the similar "Bronze" OA) has elements of both Green (there is an embargo period) and Gold (the article is published on the journal website). Piwowar, H., Priem, J., Larivière, V., et al., 2018. The state of OA: A large-scale analysis of the prevalence and impact of Open Access articles. *PeerJ*, 6, p.e4375.
7. https://creativecommons.org/faq/; also, OASPA—the Open Access Scholarly Publisher's Association—was created in 2008 and has become a "white list" for reputable publishers.
8. And a meta-analysis indicates that access leads to increased impact, as measured by citations, but confounding factors exist. McKiernan, E.C., Bourne, P.E., Brown, C.T., et al., 2016. How open science helps researchers succeed. *eLife*, 5, p.e16800.
9. Exceptions are authors based in underdeveloped countries and regions who are eligible for price reductions or waivers, see for example, https://www.research4life.org/
10. https://www.sciencemag.org/news/2017/09/are-preprints-future-biology-survival-guide-scientists

11. About 25 percent (with considerable variation between subdisciplines) of e-prints from the arXiv are never published in WoS journals. Larivière, V., Sugimoto, C.R., Macaluso, B., et al., 2014. arXiv E-prints and the journal of record: An analysis of roles and relationships. *Journal of the Association for Information Science and Technology*, 65(6), pp.1157–1169.
12. More generally this relates to the notion of the "version of record"; NISO RP-8-2008, Journal Article Versions (JAV): Recommendations of the NISO/ALPSP JAV Technical Working Group. Accessed at: https://groups.niso.org/publications/rp/RP-8-2008.pdf
13. Desjardins-Proulx, P., White, E.P., Adamson, J.J., et al., 2013. The case for open preprints in biology. *PLoS Biology*, 11(5), p.e1001563.
14. http://asapbio.org/funder-policies
15. https://peercommunityin.org/who-supports-peer-community-in/
16. Patterson, M. & Schekman, R., 2018. A new twist on peer review. *eLife*, 7, p.e36545.
17. For a recent overview, see Crotty, D. 2018. Revisiting: Six years of predatory publishing. *The Scholarly Kitchen*. Accessed at: https://scholarlykitchen.sspnet.org/2018/08/14/revisiting-six-years-predatory-publishing
18. Sorokowski, P., Kulczycki, E., Sorokowska, A., et al., 2017. Predatory journals recruit fake editor. *Nature*, 543(7646), pp.481–483.
19. Shamseer, L., Moher, D., Maduekwe, O., et al., 2017. Potential predatory and legitimate biomedical journals: Can you tell the difference? A cross-sectional comparison. *BMC Medicine*, 15(1), p.28.
20. https://beallslist.weebly.com/; https://predatoryjournals.com/journals/; https://archive.fo/9MAAD
21. Shen, C. & Björk, B.-C., 2015. "Predatory" Open Access: A longitudinal study of article volumes and market characteristics. *BMC Medicine*, 13(1), p.230.
22. An excellent resource to evaluate journal credibility is Think!Check!Submit! https://thinkchecksubmit.org/; and for OA journals the Directory of Open Access Journals (DOAJ) https://doaj.org/ and the Quality Open Access Market (QOAM) https://www.qoam.eu/

Chapter 16

1. Calcagno, V., Demoinet, E., Gollner, K., et al., 2012. Flows of research manuscripts among scientific journals reveal hidden submission patterns. *Science*, 338(6110), pp.1065–1069.
2. Paine, C.E.T. & Fox, C.W., 2018. The effectiveness of journals as arbiters of scientific impact. *Ecology and Evolution*, 8(19), pp.9566–9585.
3. Kravitz, D.J. & Baker, C.I., 2011. Toward a new model of scientific publishing: Discussion and a proposal. *Frontiers in Computational Neuroscience*, 5, p.55.
4. Larivière, V., Kiermer, V., MacCallum, C.J., et al., 2016. A simple proposal for the publication of journal citation distributions. *BioRxiv*, DOI: doi.org/10.1101/062109
5. Larivière, V. & Gingras, Y., 2010. The impact factor's Matthew effect: A natural experiment in bibliometrics. *Journal of the American Society for Information Science and Technology*, 61(2), 424–427. Cantrill S., 2016. Imperfect impact. Accessed at: https://stuartcantrill.com/2016/01/23/imperfect-impact/
6. Postma. E., 2007. Inflated impact factors? The true impact of evolutionary papers in non-evolutionary journals. *PLoS ONE*, 2(10), p.e999.
7. Aarssen, L.W., Lortie, C.J. & Budden, A.E., 2010. Judging the quality of our research: A self-assessment test. *Web Ecology*, 10(1), pp.23–26.
8. Two initiatives to promote quality choices in Open Access are the Directory of Open Access Journals (DOAJ) https://doaj.org/ and the Quality Open Access Market (QOAM) https://www.qoam.eu/
9. Van Noorden, R., 2013. Open Access: The true cost of science publishing. *Nature*, 495(7442), pp.426–429. West, J.D., Bergstrom, T. & Bergstrom, C.T., 2014. Cost effectiveness of Open Access publications. *Economic inquiry*, 52(4), pp.1315–1321. Larivière, V., Haustein, S. &

Mongeon, P., 2015. The oligopoly of academic publishers in the digital era. *PloS ONE*, 10(6), p.e0127502.
10. Crawford W. 2018. GOAJ3: Gold Open Access journals 2012–2017. Accessed at: https://waltcrawford.name/goaj3.pdf
11. Björk, B.-C. & Solomon, D., 2015. Article processing charges in OA journals: Relationship between price and quality. *Scientometrics*, 103(2), pp.373–385.
12. Bergstrom, C.T. & Bergstrom, T.C., 2004. The costs and benefits of library site licenses to academic journals. *Proceedings of the National Academy of Sciences of the United States of America*, 101(3), pp.897–902.
13. Rowlands I. & Nicholas D., 2005. New journal publishing models: The 2005 CIBER survey of journal author behaviour and attitudes. *Aslib Proceedings*, 57(6), pp.481–497.

Chapter 17

1. Bozeman, B. & Youtie, J., 2016. Trouble in paradise: Problems in academic research co-authoring. *Science and Engineering Ethics*, 22(6), pp.1717–1743.
2. Brand, A., Allen, L., Altman, M., et al., 2015. Beyond authorship: Attribution, contribution, collaboration, and credit. *Learned Publishing: Journal of the Association of Learned and Professional Society Publishers*, 28(2), pp.151–155.
3. Note that this is not true for some disciplines, for example, economics and mathematics, where authorship is often alphabetical.
4. An exception is the declaration of equal contributions, typically done among the first two or more authors on the list.
5. Sugimoto & Larivière, 2018, *op. cit.*
6. Tscharntke, T., Hochberg, M.E., Rand, T.A., et al., 2007. Author sequence and credit for contributions in multi-authored publications. *PLoS Biology*, 5(1), p.e18.
7. Note that authors are under no obligation to offer authorship to a useful reviewer report—even one that fundamentally changes the study. Reviewers assess manuscripts with the understanding that authors can use all material in the former's reports toward a revision. The usual norm is acknowledgment. Nevertheless, should authors wish to offer co-authorship to a reviewer, it is important to contact the chief editor first, explaining why.

Chapter 18

1. And indeed, some journals generally ignore cover letters. However, there is no way of knowing this since information about manuscript handling protocol is confidential.
2. Should a non-preferred reviewer be invited to assess the manuscript, then a conscientious editor will examine the submitted report in the context of the authors' original reasons for barring.
3. Verify that you have the correct journal name! It is not all uncommon that the cover letter stemming from a previously rejected manuscript mistakenly carries that journal's name.
4. https://orcid.org/. To make your work much more discoverable, ensure that you give ORCID permission to make your identifier public. You can also add a range of different outputs (not just articles) to your ORCID record.
5. Typically, an editor will invite a mix of author's and her own suggestions. This does not mean that those who actually agree to review will include your picks.
6. I suggest examining figures in the pdf carefully, and should they not be reproduced correctly, then either try rectifying the problem, or contact the editorial office for assistance.

Chapter 20

1. Whitlock, M.C., 2011. Data archiving in ecology and evolution: Best practices. *Trends in Ecology and Evolution*, 26(2), pp.61–65. Tenopir, C., Dalton, E.D., Allard, S., et al., 2015. Changes in data sharing and data reuse practices and perceptions among scientists worldwide. *PloS ONE*, 10(8), p.e0134826.
2. These are typically either institutional repositories or platforms such as ResearchGate. For lists, see http://re3Data.org; http://v2.sherpa.ac.uk/opendoar/
3. For related discussion, see Gewin, V., 2016. Data sharing: An open mind on open data. *Nature*, 529(7584), pp.117–119.
4. Stuart, D., Baynes, G., Hrynaszkiewicz, I., et al., 2018. Whitepaper: Practical challenges for researchers in data sharing. *Nature*. Accessed at: https://doi.org/10.6084/m9.figshare.5975011.v1
5. Culina, A., Baglioni, M., Crowther, T.W., et al., 2018. Navigating the unfolding open data landscape in ecology and evolution. *Nature Ecology and Evolution*, 2(3), pp.420–426.
6. Wilkinson, M.D., Dumontier, M., Aalbersberg, I.J., et al., 2016. The FAIR Guiding Principles for scientific data management and stewardship. *Scientific Data*, 3, p.160018.
7. https://www.crossref.org/; https://www.datacite.org/; www.scholix.org/; https://www.coar-repositories.org/
8. https://creativecommons.org/licenses/
9. Exclusions do exist for sensitive data, for example, patient confidentiality, endangered species locations and some proprietary data.

Chapter 21

1. The study was originally submitted to *Pan Pacific Entomologist*, but after many months without news, we learned that the chief editor had apparently quit his functions, and submitted manuscripts were irretrievable. A very unfortunate introduction into the world of publishing!
2. King, S.R., 2017. Consultative review is worth the wait. *eLife*, 6, p.e32012.
3. Nicholas, D., Watkinson, D.R., Volentine, R., et al., 2014. Trust and authority in scholarly communications in the light of the digital transition: Setting the scene for a major study. *Learned publishing: Journal of the Association of Learned and Professional Society Publishers*, 27(2), pp.121–134.
4. Note that this is likely to be biased by increased numbers of journals in the 2017 estimate, meaning that the doubling time is actually shorter.
5. Publons, 2018. Global state of peer review. Accessed at: https://publons.com/community/gspr
6. Plume, A. & van Weijen, D., 2014. Publish or perish? The rise of the fractional author…*Research Trends*, 38. Accessed at: https://www.researchtrends.com/issue-38-september-2014/publish-or-perish-the-rise-of-the-fractional-author/
7. Fox, C.W., Albert, A.Y.K. & Vines, T.H., 2017. Recruitment of reviewers is becoming harder at some journals: A test of the influence of reviewer fatigue at six journals in ecology and evolution. *Research Integrity and Peer Review*, 2, p.3. Note that the authors also found indirect evidence that some journals manage their reviewer databases to avoid repeat invites (fig. 2C), indicative that they are reacting to the tragedy. Also, the study cannot exclude the possibility that the individual reviewers censused are receiving increasing numbers of total review solicitations over time from other journals.
8. Paine & Fox, 2018, *op. cit.*
9. This is slowly changing with journals that publish peer reviews alongside articles. Ross-Hellauer, T., 2017. What is open peer review? A systematic review. *F1000Research*, 6, p.588.

10. Evidently inspired by the classic work of Garrett Hardin. Hardin, G., 1968. The tragedy of the commons. The population problem has no technical solution; it requires a fundamental extension in morality. *Science*, 162(3859), pp.1243–1248.
11. Hochberg, M.E., Chase, J.M., Gotelli, N.J., et al., 2009. The tragedy of the reviewer commons. *Ecology Letters*, 12(1), pp.2–4.
12. Fox et al., 2017, *op. cit.*
13. Donaldson, M.R., Hanson, K.C., Hasler, C.T., et al., 2010. Injecting youth into peer-review to increase its sustainability: A case study of ecology journals. *Ideas in Ecology and Evolution*, 3, pp.1–7. Hochberg, M.E., 2010. Youth and the tragedy of the reviewer commons. *Ideas in Ecology and Evolution*, 3, pp.8–10.
14. Publons, 2018, *op. cit.*
15. Fox, J. & Petchey, O.L., 2010. Pubcreds: Fixing the peer review process by "privatizing" the reviewer commons. *Bulletin of the Ecological Society of America*, 91(3), pp.325–333.
16. https://publons.com/home/
17. Ross-Hellauer, 2017, *op. cit.* Tennant, J.P., Dugan, J.M., Graziotin, D., et al., 2017. A multi-disciplinary perspective on emergent and future innovations in peer review. *F1000Research*, 6, p.1151. Polka, J.K., Kiley, R., Konforti, B., et al., 2018. Publish peer reviews. *Nature*, 560(7720), pp.545–547. Parker, T.H., Griffith, S.C., Bronstein, J.L., et al., 2018. Empowering peer reviewers with a checklist to improve transparency. *Nature Ecology and Evolution*, 2(6), pp.929–935.
18. The Royal Society has recently introduced open peer review at two of its journals. https://blogs.royalsociety.org/publishing/publication-of-open-peer-review/
19. Another more recently proposed form of post-publication feedback is open, interoperable annotations. Naydenov, A. 2018. Guest post: The time for open and interoperable annotation is now. *The Scholarly Kitchen*. Accessed at: https://scholarlykitchen.sspnet.org/2018/08/28/all-about-open-annotation/

Chapter 22

1. Several major institutes have recently cancelled "Big deals," having failed to negotiate satisfactory terms. See https://sparcopen.org/our-work/big-deal-cancellation-tracking/
2. Poynder, R. 2018. The Open Access Big deal: Back to the future. Accessed at: https://poynder.blogspot.com/2018/03/the-open-access-big-deal-back-to-future.html
3. https://www.research4life.org/
4. Pinfield, S., Salter, J. & Bath, P.A., 2016. The "total cost of publication" in a hybrid open-access environment: Institutional approaches to funding journal article-processing charges in combination with subscriptions. *Journal of the Association for Information Science and Technology*, 67(7), pp.1751–1766.
5. Johnson et al., 2018, *op. cit.*
6. For projections see 2019 EBSCO Serials Price Projection Report. Accessed at: https://www.ebscohost.com/promoMaterials/2019_Serials_Price_Projections.doc.pdf
7. Geschuhn, K. & Stone, G., 2017. It's the workflows, stupid! What is required to make "offsetting" work for the Open Access transition? *Insights: The UKSG Journal*, 30(3), pp.103–114.
8. There is some evidence that authors based at lower-ranked universities tend to choose non-APC as opposed to APC journals. Siler, K., Haustein, S., Smith, E., et al., 2018. Authorial and institutional stratification in Open Access publishing: The case of global health research. *PeerJ*, 6, p.e4269.
9. Else, H., 2018. Radical open-access plan could spell end to journal subscriptions. *Nature*, 561(7721), pp.17–18. Funders also want a more transparent cost structure and are committed to changing the research evaluation system, using DORA as a starting point.
10. See also https://oa2020.org/

11. A recent SpringerNature whitepaper argues that "The additional cost/time/risk/disruption for the whole research ecosystem as well as to publishers would be huge compared with the opportunity to progress an orderly evolution." Lucraft, M., Draux, H. & Walker, J. (2018) Assessing the Open Access effect in hybrid journals. Accessed at: https://doi.org/10.6084/m9.figshare.6396290
12. Funder mandates are likely to increasingly drive OA publishing behavior; Larivière, V. & Sugimoto, C.R., 2018. Do authors comply when funders enforce open access to research? *Nature*, 562(7728), pp.483–486.
13. Johnson et al., 2018, *op. cit.*
14. Moreover, these surpluses do not factor in membership proceeds for society journals.
15. Solomon, D.J. & Björk, B.-C., 2012. A study of Open Access journals using article processing charges. *Journal of the American Society for Information Science and Technology*, 63(8), pp.1485–1495. Van Noorden 2013, *op. cit.*
16. Larivière et al., 2015, *op. cit.*
17. Bergstrom & Bergstrom, 2004, *op. cit.*. Dewatripont, M., Ginsburgh, V., Legros, P., et al., 2007. Pricing of scientific journals and market power. *Journal of the European Economic Association*, 5(2–3), pp.400–410. Bergstrom, T.C., Courant, P.N., McAfee, P., et al., 2014. Evaluating big deal journal bundles. *Proceedings of the National Academy of Sciences of the United States of America*, 111(26), pp.9425–9430. Solomon & Björk, 2012, *op. cit.*
18. Exceptions exist. See for example, Patterson, M. & McLennan, J. Inside *eLife*: What it costs to publish. Accessed at: https://elifesciences.org/inside-elife/a058ec77/what-it-costs-to-publish
19. Johnson et al., 2018, *op. cit.*
20. Houghton, J., Rasmussen, B., Sheehan, P., et al., 2009. *Economic Implications of Alternative Scholarly Publishing Models: Exploring the Costs and Benefits*. Joint Information Systems Committee (JISC), London and Bristol. Van Noorden 2013, *op. cit.*
21. Solomon & Björk, 2012, *op. cit.*; Van Noorden 2013, *op. cit.*
22. Though not completely. See Corbyn, Z., 2013. Price doesn't always buy prestige in Open Access. *Nature*. Accessed at: www.nature.com/doifinder/10.1038/nature.2013.12259
23. Fyfe, A., Kelly, C., Curry, S., et al., 2017. Untangling academic publishing: A history of the relationship between commercial interests, academic prestige and the circulation of research. *Zenodo*, p.546100.
24. Björk, B. & Solomon, D. (2014) *Developing an Effective Market for Open Access Article Processing Charges*. Wellcome Trust, London. Björk, B.-C., 2017. Scholarly journal publishing in transition—from restricted to Open Access. *Electronic Markets*, 27(2), pp.101–109.
25. Dewatripont et al., 2007, *op. cit.*; Solomon & Björk, 2012, *op. cit.*; West et al. 2014, *op. cit.*
26. Bergstrom et al., 2014, *op. cit.*
27. European Commission, 2006. Study on the economic and technical evolution of the scientific publication markets in Europe. Accessed at: https://ec.europa.eu/research/openscience/pdf/openaccess/librarians_2006_scientific_pub_study.pdf
28. The APC per article at the most profitable publishers would likely exceed tens of thousands of US$. In April 2004 the Nature Publishing Group estimated that they would need to charge approximately £30,000 per published article to match what they receive per article in annual subscriptions; Accessed at: https://publications.parliament.uk/pa/cm200304/cmselect/cmsctech/399/399we163.htm
29. As of 2013, five publishers accounted for 50 percent of all published articles referenced by the Web of Science. Larivière et al., 2015, *op. cit.*
30. The issues extend beyond publishing *per se* to control over data storage and analytics. Posada, A., & Chen, G., 2017. Publishers are increasingly in control of scholarly infrastructure and why we should care. The Knowledge G.A.P. Geopolitics of Academic Production. Accessed at: http://knowledgegap.org/index.php/sub-projects/rent-seeking-and-financialization-of-the-academic-publishing-industry/preliminary-findings/

31. A recent analysis suggests that large research groups are more likely to develop science, whereas small (less-funded) groups disrupt science through new ideas. This analysis did not investigate likelihood of these groups publishing in APC journals. Wu, L., Wang, D. & Evans, J.A., 2019. Large teams develop and small teams disrupt science and technology. Nature, 566(7744), pp.378–382.
32. Johnson, R., Fosci, M., Chiarelli, A., et al. 2017. Towards a competitive and sustainable OA market in Europe—a study of the Open Access market and policy environment. *Zenodo*, p.401029.

Chapter 23

1. Impact as used in this book refers to how a published article changes science, as measured by citation metrics. A more inclusive notion of impact extends to the influence of tools and information (e.g. software, datasets, blogs, podcasts, etc.) on areas such as technology, medicine, agriculture and species conservation. Open Science promises to integrate this broader concept of impact (e.g. McKiernan et al., 2016, *op. cit.*).
2. The Semantic scholar https://www.semanticscholar.org/ aims to account for the context of a citation, and multiple citations of the same paper.
3. Garfield, E., 1972. Citation analysis as a tool in journal evaluation: Journals can be ranked by frequency and impact of citations for science policy studies. *Science*, 178(4060), pp.471–479.
4. For historical account of impact factors, see Archambault, É. & Larivière, V., 2009. History of the journal impact factor: Contingencies and consequences. *Scientometrics*, 79(3), pp.635–649.
5. The official JIF is produced by the Web of Science, owned by Clarivate Analytics.
6. https://clarivate.com/essays/journal-selection-process/
7. Hirsch, J.E., 2005. An index to quantify an individual's scientific research output. *Proceedings of the National Academy of Sciences*, 102(46), pp.16569–16572.
8. Yong, A., 2014. A critique of Hirsch's citation index: A combinatorial Fermi problem. *Notices of the American Mathematical Society. American Mathematical Society*, 61(09), p.1040.
9. Bollen, J., van de Sompel, H., Hagberg, A., et al., 2009. A principal component analysis of 39 scientific impact measures. *PloS ONE*, 4(6), p.e6022. Sugimoto & Larivière, 2018, *op. cit.* See periodic table of scientometric indicators www.elprofesionaldelainformacion.com/notas/wp-content/uploads/2018/06/tablaper3.pdf
10. Chen, P., Xie, H., Maslov, S., et al. 2006. Finding scientific gems with Google. *arXiv*, p.0604130v1. Ding, Y., Yan, E., Frazho, A., et al., 2009. PageRank for ranking authors in co-citation networks. *Journal of the American Society for Information Science and Technology*, 60(11), pp.2229–2243.
11. https://www.altmetric.com/
12. Casadevall, A. & Fang, F.C., 2014. Causes for the persistence of impact factor mania. *mBio*, 5(2), pp.e00064–e00014.
13. Wilsdon, J., Allen, L., Belfiore, E., et al. 2015. The metric tide: Report of the independent review of the role of metrics in research assessment and management. Higher Education Funding Council for England Report. Accessed at: http://eprints.whiterose.ac.uk/117033/1/2015_metric_tide.pdf. For reward structure in China, see Quan, W., Chen, B. & Shu, F., 2017. Publish or impoverish: An investigation of the monetary reward system of science in China (1999–2016). *arXiv*, p.1707.01162.
14. Hochberg, M., 2014. Good science depends on good peer review. *Ideas in Ecology and Evolution*, 7(1), pp.77–83.
15. Daniel, B., 2015. Big Data and analytics in higher education: Opportunities and challenges: The value of Big Data in higher education. *British Journal of Educational Technology: Journal of the Council for Educational Technology*, 46(5), pp.904–920.
16. http://eigenfactor.org/projects/posts/citescore.php

17. Larivière et al., 2016, *op. cit.*; Sugimoto & Larivière, 2018, *op. cit.*
18. Brembs, B., Button, K. & Munafò, M., 2013. Deep impact: Unintended consequences of journal rank. *Frontiers in Human Neuroscience*, 7, p.291.
19. Seglen, P.O., 1997. Why the impact factor of journals should not be used for evaluating research. *BMJ*, 314(7079), pp.497–497.
20. This refers to a small number of questionable examples. Placing accepted manuscripts online early is an entirely legitimate way of publishing studies without delay.
21. Brembs, B., 2018. Prestigious science journals struggle to reach even average reliability. *Frontiers in Human Neuroscience*, 12, p.37.
22. Neff, B.D. & Olden, J.D., 2010. Not so fast: Inflation in impact factors contributes to apparent improvements in journal quality. *Bioscience*, 60(6), pp.455–459.
23. Sugimoto & Larivière, 2018, *op. cit.*
24. Althouse, B.M., West, J.D., Bergstrom, C.T., et al., 2009. Differences in impact factor across fields and over time. *Journal of the American Society for Information Science and Technology*, 60(1), pp.27–34. Larivière, V. & Sugimoto, C.R., 2018. The journal impact factor: A brief history, critique, and discussion of adverse effects. Forthcoming in Glanzel, W., Moed, H.F., Schmoch, U., et al. *Springer Handbook of Science and Technology Indicators*. Springer International Publishing, Cham, Switzerland. Ioannidis, J.P.A., 2006. Concentration of the most-cited papers in the scientific literature: Analysis of journal ecosystems. *PloS ONE*, 1, p.e5.
25. Hutchins, B.I., Yuan, X., Anderson, J.M., et al., 2016. Relative Citation Ratio (RCR): A new metric that uses citation rates to measure influence at the article level. *PLoS Biology*, 14(9), p.e1002541.
26. https://sfdora.org/. Hicks, D., Wouters, P., Waltman, L., et al., 2015. Bibliometrics: The Leiden Manifesto for research metrics. *Nature*, 520(7548), pp.429–431. Wilsdon et al., 2015, *op. cit.*
27. Curry, S., 2018. Let's move beyond the rhetoric: It's time to change how we judge research. *Nature*, 554(7691), pp.147–147. Moher, D., Naudet, F., Cristea, I.A. et al., 2018. Assessing scientists for hiring, promotion, and tenure. *PLoS Biology*, 16(3), p.e2004089. Hicks et al., 2015,*op. cit.*

Chapter 24

1. Bornmann & Mutz, 2015, *op. cit.*
2. Johnson et al., 2018, *op. cit.*
3. Sugimoto & Larivière, 2018, *op. cit.*
4. Rowlands, I., Clark, D., Jamali, H., et al., 2012. PEER D5.2 USAGE STUDY Descriptive statistics for the period March to August 2011. Accessed at: https://hal.inria.fr/hal-00738501
5. Abt, H.A., 1998. Why some papers have long citation lifetimes. *Nature*, 395(6704), pp.756–757.
6. Van Noorden, R., 2017. The science that's never been cited. *Nature*, 552(7684), pp.162–164.
7. Merton, R.K., 1968. The Matthew effect in science. The reward and communication systems of science are considered. *Science*, 159(3810), pp.56–63.
8. Wang, D., Song, C. & Barabási, A.-L., 2013. Quantifying long-term scientific impact. *Science*, 342(6154), pp.127–132.
9. Note that this is an idea. I know of no journals that actually advertise citation appendices in their *advice to authors*.

Chapter 25

1. Nicholas D., Abrizah, A., Boukacem-Zeghmouri, C., et al., 2018. Early-career researchers: The harbingers of change. Final report. CIBER Research Ltd. Accessed at: http://ciber-research.eu/download/20181218-Harbingers3_Final_Report-Nov2018.pdf

2. For career biographies and advice, see Women's contribution to basic and applied evolutionary biology. *Evolutionary Applications*, 2016, 9(1), pp.1–310. Nettle, D. 2017. Hanging on the edges: staying in the game? Accessed at: https://www.danielnettle.org.uk/wp-content/uploads/2017/09/Staying-in-the-game.pdf. Campana, S.E. 2018. Twelve easy steps to embrace or avoid scientific petrification: Lessons learned from a career in otolith research. *ICES Journal of Marine Science: Journal du Conseil*, 75(1), pp.22–29.
3. However, it is important to realize that we learn from failed research—and indeed sometimes we learn *more* from failure than from a project that seamlessly produces positive results and publications. Clearly, we ultimately need to complete and publish a study. The more usual way to learn from shortcomings or failure is through pilot studies.
4. Simon, H.A. & Chase, W.G., 1973. Skill in chess. *American Scientist* 61(4), pp.394–403.
5. This said, some papers are scientifically so well-conducted, interesting and important that it almost makes no difference where they are published.

Chapter 26

1. Evidence for the effect of collaboration on article impact varies between disciplines; see Leimu, R. & Koricheva, J., 2005. Does scientific collaboration increase the impact of ecological articles? *Bioscience*, 55(5), p.438. Franceschet, M. & Costantini, A., 2010. The effect of scholar collaboration on impact and quality of academic papers. *Journal of Informetrics*, 4(4), pp.540–553.
2. Simms, A. & Nichols, T., 2014. Social loafing: A review of the literature. *Journal of Management Policy and Practice*, 15(1), pp.58–67.

Chapter 28

1. See Tennant et al., 2017, *op. cit.* for overview of the history of peer review.
2. PRC 2016, Publishing Research Consortium peer review Survey 2015 Report, Mark Ware Consulting. Accessed at: http://publishingresearchconsortium.com/index.php/prc-documents/prc-research-projects/57-prc-peer-review-survey-2015/file
3. Publons, 2018. *op. cit.*
4. It is generally not good practice to agree to review a paper and then decide that you are not qualified. Contact the editorial office if you have questions about your suitability to conduct a review.
5. At *Ecology Letters*, personal comments were addressed by contacting the reviewer and requesting that the report be rewritten in an impersonal, scientific way.
6. McPeek, M.A., DeAngelis, D.L., Shaw, R.G., et al., 2009. The golden rule of reviewing. *The American Naturalist*, 173(5), pp.E155–E158.
7. See interesting interviews in Yoder J., 2014. Why we don't sign our peer reviews. Accessed at: www.molecularecologist.com/2014/04/why-we-dont-sign/
8. Another possible issue with signing a report is that many journals routinely send the decision letter together with reviews to all reviewers. If the journal is remiss (as once happened to me, and that's how I know this can occur) and omits to remove your name from your report, then the other reviewers will see your identity.
9. Unless you have reasons for wanting the authors to know who you are (and eventually contacting you), there is no reason (beyond responsibility, which can be achieved by signing and removing your name just before submitting) to sign your report.
10. https://publons.com/community/academy/; http://www.peere.org/

11. Receiving reviews *also* carries responsibility. Reviews and reviewer identities are confidential material between the reviewer, the journal and authors. Authors can, of course, thank signed reviewers in the acknowledgments, but should never mention the reviewer's name to any third party.
12. Lee, C.J., Sugimoto, C.R., Zhang, G., et al., 2013. Bias in peer review. *Journal of the American Society for Information Science and Technology*, 64(1), pp.2–17. Tomkins, A., Zhang, M. & Heavlin, W.D., 2017. Reviewer bias in single- versus double-blind peer review. *Proceedings of the National Academy of Sciences of the United States of America*, 114(48), pp.12708–12713.
13. Enserink, M., 2017. Few authors choose anonymous peer review, massive study of *Nature* journals shows. *Science*. Accessed at: www.sciencemag.org/news/2017/09/few-authors-choose-anonymous-peer-review-massive-study-nature-journals-shows.

Chapter 29

1. For overviews, see Bik, H.M. & Goldstein, M.C., 2013. An introduction to social media for scientists. *PLoS Biology*, 11(4), p.e1001535. Van Noorden, R., 2014. On-line collaboration: Scientists and the social network. *Nature* 512(7514), pp.126–129.
2. Sugimoto, C.R., Work, S., Larivière, V., et al., 2017. Scholarly use of social media and altmetrics: A review of the literature. *Journal of the Association for Information Science and Technology*, 68(9), pp.2037–2062.
3. Evidence is mixed that social media interest correlates with citations (Sugimoto et al., 2017, *op. cit.*).
4. Sugimoto et al., 2017, *op. cit.*
5. Schnitzler, K., Davies, N., Ross, F., et al., 2016. Using Twitter™ to drive research impact: A discussion of strategies, opportunities and challenges. *International Journal of Nursing Studies*, 59, pp.15–26.
6. For example, *Slideshare* (https://www.slideshare.net/), hosts slideshows, webinars and videos.
7. Some of these and other social collaboration networks are currently confronted with legal issues of posting copyright-protected documents. Jamali, H.R., 2017. Copyright compliance and infringement in ResearchGate full-text journal articles. *Scientometrics*, 112(1), pp.241–254.
8. For this and related discussion, see West, J., Bergstrom, T. & Bergstrom, C.T., 2010. Big Macs and Eigenfactor scores: Don't let correlation coefficients fool you. *Journal of the American Society for Information Science and Technology*, 61(9), pp.1800–1807.
9. de Winter, J.C.F., 2015. The relationship between tweets, citations, and article views for PLoS ONE articles. *Scientometrics*, 102(2), pp.1773–1779; and reference therein.
10. Sugimoto et al., 2017, *op. cit.*

Chapter 30

1. For recent review see LERU, 2018. Open Science and its role in universities: a roadmap for cultural change. Advice paper 24. Accessed at: https://www.leru.org/files/LERU-AP24-Open-Science-full-paper.pdf
2. Johnson et al., 2017. *op. cit.*
3. Piwowar et al., 2018. *op. cit.*
4. Crotty, D., 2016. Can highly selective journals survive on APCs? *The Scholarly Kitchen*. Accessed at: https://scholarlykitchen.sspnet.org/2016/10/10/can-highly-selective-high-end-journals-survive-on-apcs/
5. Eisen, M., 2015. The inevitable failure of parasitic green Open Access. Accessed at: www.michaeleisen.org/blog/?p=1710

6. Guédon, J.-C., Jubb, M., Kramer, B., et al. 2019. Future of scholarly publishing and scholarly communication: Report of the Expert Group to the European Commission. Accessed at https://haris.hanken.fi/portal/files/10149067/KI0518070ENN.en.pdf
7. Guédon, 2017, *op. cit.*; Johnson et al. 2018, *op. cit.*; for possible issues with funder OA platforms see Ross-Hellauer, T., Schmidt, B. & Kramer, B., 2018. Are funder Open Access platforms a good idea? *SAGE Open*, 8(4), p.215824401881671.
8. See e.g., for funders: http://v2.sherpa.ac.uk/view/funder_visualisations/1.html; institutional policies: http://roarmap.eprints.org/view/policymaker_type/research=5Forg.html; national policies: https://www.openaire.eu/frontpage/country-pages
9. https://tspoa.org/
10. Jubb, M., Plume, A., Oeben, S., et al., 2017. Monitoring the transition to Open Access. Report. Universities UK, London.
11. Rabesandratana, T., 2019. The world debates open-access mandates. *Science*, 363(6422), pp.11–12. See following exchange: Burgman, M., 2018. Open Access and academic imperialism. *Conservation Biology* DOI: 10.1111/cobi.13248. Lewis, D.W. 2018. Conservation biology as an example of the dilemmas facing scholarly society publishing. Accessed at: https://scholarworks.iupui.edu/handle/1805/18026
12. Ioannidis, J.P.A., 2018. Meta-research: Why research on research matters. *PLoS Biology*, 16(3), p.e2005468. "Science of science" is a similar approach, but is more oriented toward the fundamental understanding of pattern and process in science; see Fortunato, S., Bergstrom, C.T., Börner, K., et al., 2018. Science of science. *Science*, 359(6379), p:eaao0185.

Suggested Reading

Grammar, Style and Writing Strategy

Steven Pinker. 2014. *The Sense of Style. The Thinking Person's Guide to Writing in the 21st Century*. Penguin Books, New York.

David Allen. 2015. *Getting Things Done. The Art of Stress-free Productivity*. Penguin Books, New York.

General Writing and Publishing

Joshua Schimel. 2012. *Writing Science. How to Write Papers That Get Cited and Proposals That Get Funded*. Oxford University Press, Oxford, UK.

Janice R. Matthews and Robert W. Matthews. 2014. *Successful Scientific Writing. A Step-By-Step Guide for the Biological and Medical Sciences*. 4th edition. Cambridge University Press, Cambridge, UK.

Robert A. Day and Barbara Gastel. 2016. *How to Write and Publish a Scientific Paper*. 8th edition. Greenwood Press, CT.

Open Science and Open Access

Peter Suber. 2012. *Open Access*. The MIT Press, Cambridge, MA.

Scholarly Communication

Brook Borel. 2016. *The Chicago Guide to Fact-Checking*. University of Chicago Press, Chicago, IL.

Rick Anderson. 2018. *Scholarly Communication. What Everyone Needs to Know*. Oxford University Press, Oxford, UK.

Peer Review

Irene Hames. 2007. *Peer Review and Manuscript Management in Scientific Journals. Guidelines for Good Practice*. Blackwell Publishing, Oxford, UK.

Publons. 2018. Global State of Peer Review. https://publons.com/community/gspr

Science and Being a Scientist

Corey J. A. Bradshaw. 2018. *The Effective Scientist: A Handy Guide to a Successful Academic Career*. Cambridge University Press, Cambridge, UK.

David K. Hull. 1988. *Science as a Process: An Evolutionary Account of the Social and Conceptual Development of Science*. Chicago, IL: University of Chicago Press, USA.

Index

A
abstracts, *60*, 63–6
acknowledgment *vs.* authorship, 126
altmetrics, 177
appeals, 103, 144–5
Article Processing Charge (APC), 106–7, 120, 166–71, 225
authors, 84, 156
　editorial office replying to queries of, 102
　expectations from journals, 85–6
　integrity, 187
　responsibility, 17, 18
authorship, *125*
　author order, 129–31
　contributions meriting, 126
　disputes, avoiding, 132
　justification, 127–9
　notification of, 131–2
　responsibility and accountability, 129

C
cascade journals, 162–63
chief editor, 83
　publication decisions, 92–3
　role in evaluation process, 93–4
citation, *13*
　biases, 16, 17, 181, 187
　criteria for, 14–16
　and impact, 117
citation metrics, *172*
　citation half-life, 177
　h index, 175–6
　i10 index, 177
　immediacy index, 177
　issues with, 179–81
　journal impact factor: *see* journal impact factor
　total citations, 175–6
　5-year impact factor, 177
collaborative studies, *199*
　advantages of 202
　division of labor in, 204
　factors influencing success of, 203
　vs. solo author, 201
collaborative writing, 39–40
conflicts of interest, 212
copyright, 150, 166–8
　infringement, 23–4

cost of publishing, *164*
　factors contributing to, 166
　hybrid OA journal, 167–8
　open access, 166, 168
　scholarly journals, 167
　subscriptions, 169
cover letter, 93–4, *133*
　elements of, 135–6
　importance of, 134–5

D
data archiving, *146*
　and conservation, 147
　adopting policy of, 148
　advantages and concerns, 148–9
　for data reanalysis, 149
　sharing, 147, 149–50
decision letter, *139*
　for acceptance, 141
　appeals, 144–5
　for rejection, 140
　replying to, 141–4
　for revision, 140–1
decision time, 118–19
desk rejection, 92, 116, 159–60, 192
discussion section, 37, 51–9
　first paragraph of, 56
double-blind review, 217

E
edited volumes, publishing in, 112–13
editorial board, 83–4, 94
editorial decision: *see* publication decision
editorial office, 82–3, 89, 99–100, 102
emphasis
　terminology and jargon, 74–5
　through wording, 76
　visual material as means of, 77
evaluation culture, 11, 110, 171

F
figures, *60*, 66–7, 77

G
gender bias, 85
gold OA, 106–7, 167–8
　pricing, 120, 170

green OA, 106–7, 167–168
   data archiving, 148

H
handling editor, 83, 88, 101
hook, 64–5
hybrid journals, 106, 167
   growth of, 168
   pricing, 167–8

I
impact, 8, 10–11, 173, 177–9
   factors for research fields, 181
   and influence, 177
   issues, 179–181
   and journals, 85
   and journal choice, 115–7
impact factor. *See* journal impact factor
IMRaD, *50*, 51–2
influence, 177–8
introduction section, 37, 50–2
   components, 52–3
   in *Models*, 45–8
   paragraph structure in, 53

J
jargon and terminology, 74–5
journal, 10–11
   and plagiarism, 23–4
   choice. *See* journal selection
   decision and publication times, 100
   expectations from authors, 85–6
   functions, 81–2
   reputation, 119–20
   selectivity, models of, 86–7
   workflow, 82–3
journal impact factor, 116–17, 173–4
   and evaluation culture, 110
   calculation, 173–4
journal rank, 175
   journal operations, *81*
   chief editor, 83, 92–4
   editorial board, 83
   editorial office, 83, 93
   handling editor/editor, 83, 88, 93, 94
   reviewers: *see* peer review, *see* reviewers
journal selection, criteria for, *114*
   decision time, 118–19
   history of publication, 120–1
   journal impact factor, 116–17
   mixed, 121
   prices to authors and subscribers, 120

   rapid publication decisions, 119
   reputation, 119–20

L
literature search *13*
   ascending and descending, 21
   databases, 20–1
   use of keywords, 20–1

M
meta-research, 226–7
methods section, 36–7, 51–2
Models, *42*, 63, 224
   choosing, 43–4
   importance in scientific writing, 43
   plagiarism and, 44–5
   motivation for writing: goal achievement, 28–9
   reactive and proactive, 28

N
notification of authorship, 131–2

O
Open Access (OA), 106–7, 120, 166, 168
   factors influencing shift towards, 225–6
open data
   preprints, 107–11
   see data archiving
Open Science, 149, 227
outlining, 35–7

P
paragraph structure, 52–3
peer review, 9, 84, 86, *153*
   aims of, 210
   alternatives to, 154
   benefits of, 216–17
   challenge of, 87–90
   ethics, 217
   the tragedy of the reviewer commons, 157–63
Plan-S, 168, 171, 226
predatory journals, 111–12, 171
preprints, 107–11
platforms, 154
productivity of scientific writing, 6,8
   attaining career goals through, 12
   high, drawbacks of, 9
   and quality, link between, 8–10
   publication decisions, *139*, 140–41
   timeliness of, 118–9

Q

quality of scientific writing, 6,7–8
  and productivity, link between, 8–10

R

reading, scientific literature, *13*
  importance in writing paper, 18
  strategy, 18–19
reproducibility crisis, 9, 227
results section, 37, *50*, 51–2, 54
reviewers, 84, 88, 98, 100, 155
  anonymity, 215–16
  assessments, 86
  challenges with, 87–90
  comments and critiques on paper, 94
  difficulty in getting, 156
  letter replying to, 142–4
  over-solicitation of, 157–8
  ways to acknowledge and reward, 161–2
  see also peer review
reviewing, *209*
  etiquette in, 212–14
  guidelines in structuring report, 215–16
  importance of, 210
revision: carefully verifying, 215
  decision letter, 140–1
  in writing manuscripts, 38–9
  recommendation for, 88
  of rejected manuscripts, 162–3

S

social media, *218*
  access to information, 219
  blogs, 221
  engaging interactions among stakeholders, 220
  influence on communication, 218–20, 222
  limitations of, 220–1
storytelling, 33, 34–5
style, writing, 73
  article types and, 75
  qualifiers, 73–4
  terminology and jargon, 74–5
subscription journals, 105–6, 120, 165–6
  and paywalls, 20
  pricing, 169–70

T

titles, *60*
  dimensions in constructing, 62–3
  importance of, 61

W

writing time, 38
  planning, 33–4
  and quality, link between, 33

Made in the USA
Monee, IL
04 May 2026

49438119R00162